Economic Com

Series Editors

Uwe Cantner, Dept of Econ & Business Administration, Friedrich Schiller University Jena, Jena, Germany

Kurt Dopfer, St. Gallen, Switzerland

John Foster, University of Queensland, Brisbane, Australia

Andreas Pyka, University of Hohenheim, Stuttgart, Germany

Paolo Saviotti, INRA GAEL, Université Pierre Mendès-France, Grenoble, France

Research on the dynamics and evolution of economies and their subsystems has increasingly attracted attention during the past three decades. Micro-level activities related to innovation, imitation, adoption and adaptation are sources of often important impulses for abrupt as well as continuous, local as well as generic, systemic as well as modular changes in mesoeconomic and macroeconomic structures. These changes driven by selection and co-evolution, by path-dependencies and lock-ins as well as by transitions and jumps in turn feedback to the system's microeconomic units. Out of this a highly interconnected system of interaction between agents, subsystems and levels of aggregation emerges and confronts researchers with the challenges to grasp, reduce and understand the complexity and her dynamics involved. The series "Economic Complexity and Evolution" addresses this wide field of economic and social phenomena and attempts to publish work which comprehensively discusses special topics therein allowing for a broad spectrum of different methodological viewpoints and approaches.

All titles in this series are peer-reviewed.

This book series is indexed in Scopus.

More information about this series at http://www.springer.com/series/11583

Michael P. Schlaile
Editor

Memetics and Evolutionary Economics

To Boldly Go Where no Meme has Gone Before

Springer

Editor
Michael P. Schlaile
Institute of Education, Labor and Society (560)
and Institute of Economics (520)
University of Hohenheim
Stuttgart, Germany

Dissertation University of Hohenheim, 2019, D100
Stuttgart, Germany

ISSN 2199-3173 ISSN 2199-3181 (electronic)
Economic Complexity and Evolution
ISBN 978-3-030-59957-7 ISBN 978-3-030-59955-3 (eBook)
https://doi.org/10.1007/978-3-030-59955-3

© The Editor(s) (if applicable) and The Author(s), under exclusive license to Springer Nature Switzerland AG 2021, corrected publication 2021
This work is subject to copyright. All rights are solely and exclusively licensed by the Publisher, whether the whole or part of the material is concerned, specifically the rights of translation, reprinting, reuse of illustrations, recitation, broadcasting, reproduction on microfilms or in any other physical way, and transmission or information storage and retrieval, electronic adaptation, computer software, or by similar or dissimilar methodology now known or hereafter developed.
The use of general descriptive names, registered names, trademarks, service marks, etc. in this publication does not imply, even in the absence of a specific statement, that such names are exempt from the relevant protective laws and regulations and therefore free for general use.
The publisher, the authors and the editors are safe to assume that the advice and information in this book are believed to be true and accurate at the date of publication. Neither the publisher nor the authors or the editors give a warranty, expressed or implied, with respect to the material contained herein or for any errors or omissions that may have been made. The publisher remains neutral with regard to jurisdictional claims in published maps and institutional affiliations.

This Springer imprint is published by the registered company Springer Nature Switzerland AG
The registered company address is: Gewerbestrasse 11, 6330 Cham, Switzerland

Foreword

The core concept of neoclassical mainstream economics, which ultimately represents an adaptation of Newton's classical mechanics to the economic system, has long been criticized; not least because the classical model of physics has already been outdated for about a hundred years (since Einstein's two theories of relativity and quantum physics) and thus the question arises whether the neoclassical market mechanics can be a model that is able to adequately represent the economic world. In addition to other concepts (e.g., *behavioral economics*), *evolutionary economics* also tries to develop scientifically fruitful proposals for a better description and analysis of the economic and social processes in a thoroughly evolutionary world with all its innovations (keyword: *innovation economics*), replications, and the disappearance of the old. This is a far more complex challenge than merely mapping a stereotypical market mechanism.

Years ago, the evolutionary biologist Richard Dawkins made the proposal to assume a replicator of social or cultural evolution, which he called the "meme" (in analogy to the replicator of biological evolution, the "gene"). "Memes" are replicators of human culture, which reproduce themselves as units of information, imitation, and instruction, but which also develop and evolve creatively, thus advancing social and especially economic evolutionary processes. These mental information units are the content of cultural and economic evolution, and they are becoming concrete in habits, which are replicating and updating this information.

Michael Schlaile takes up this proposal by developing his own economic concept of *economemetics*. Against the background that imitation plays an important role in social or economic systems and that the mechanisms of cultural evolutionary processes are still underrepresented in economic analysis, the author uses the term "economemetics ... to describe this highly interdisciplinary research endeavor that aims to explore the potential of utilizing memetic approaches for (evolutionary) economics" (quoted from Chap. 3). The combination of the terms "memetics" and "econo" signals his goal of synergizing memetics and (evolutionary) economics or integrating memetics with evolutionary economics.

In Chap. 7, Michael Schlaile also discusses fundamental questions raised by the concept of memetics in general and economemetics in particular:

- Firstly, there is the question to what extent memes can be regarded as "agents" that are able (or unable) to infect human thinking actively and almost intentionally. Michael Schlaile proposes a solution to the problem by using the concepts of "actants", "affordances", and "quasi-agents".
- The second problem revolves around the relationship between replication (or: "remix"), on the one hand, and "creativity" and "novelty" in the evolutionary processes of society and economy, on the other hand.
- And finally, there is the issue of "memethics", i.e., the normative implications of (econo)memetics. There's no doubt that "moral memes" belong to the realm of cultural evolution, which shape the evolution of moral or normative knowledge and what Dawkins has called the "changing moral *Zeitgeist*".

With his "economemetics", Michael Schlaile is presenting an integrative concept that is both interdisciplinary in its approach and innovative for economic theory. This research originates in two different environments, namely ethics and economic dynamics, which offered Michael Schlaile a prolific terrain for exploring the opportunities for interdisciplinary research. The book at hand proves that the author mastered this challenge in an excellent way and opens new research trajectories, which might have an impact on the future of our disciplines. For sure, Michael Schlaile showed that new insights can be achieved by taking seriously the mission to go beyond tight disciplinary borders and to allow for cross-fertilization between distinct intellectual traditions.

Stuttgart, Germany   Michael Schramm
                                                                                     Andreas Pyka

Preface and Acknowledgments

This work is the result of a journey that took approximately seven years, which occasionally felt like a smooth train ride toward comprehension and more frequently like an intellectual ride on the academic rollercoaster. Neither a train nor a rollercoaster can operate with just one passenger and the same obviously holds true for this book.

Without the continuous support, encouragement, and also trust of my doctoral advisor Prof. Dr. Michael Schramm and my second supervisor Prof. Dr. Andreas Pyka this book would not have been possible. In this connection, I would also like to express my deep gratitude to Prof. Dr. Bernd Ebersberger for having served as the chair of the examination committee and for his ongoing guidance and advice. Moreover, since this book contains a collection of articles that are coauthored with other researchers, colleagues, and friends, I am deeply indebted to Dr. Kristina Bogner, Theresa Knausberg, Laura Mülder, Dr. Matthias Müller, and Johannes Zeman. Special thanks are due to Johannes, Kristina, and Matthias for their patience, for sharing their technical know-how, and for helping me to translate ideas into computational models. During the past seven years, I also had the pleasure of working on stimulating projects and publications (often closely related to the topics of this book) with the following colleagues, with whom I had inspiring discussions that shaped the evolution of my thoughts: Wolfgang Böck, Joe Brewer, Dr. Marcus Ehrenberger, Katharina Klein, and especially my "office roomie" Dr. Sophie Urmetzer.

This book would also not have been possible without the support and assistance of various other coworkers and research assistants at the University of Hohenheim: Thank you, Dr. Andreas Diße, Katharina Eckstein, Jessica Gruber, Claudia Hauser, Bianca Janic, Dr. Helen Mengis, Sinan Qazzazie, Anna-Lena Strobel, Alina Thum, Michael Volz, Dr. Christoph Wagner, Patrik Walter, and Lukas Zuschrott.

I am also grateful to numerous other colleagues and collaborators, conference participants, and journal reviewers for comments and criticism. These people are explicitly named in the acknowledgments of the respective articles (Chaps. 3–6).

Finally, I want to thank my friends and my family for enduring my moods, my absent-mindedness, and for supporting me in every possible way. Special thanks are due to my aunts Sigrid Bareiss and Dr. Christiane Pergande-Bareiss, my uncle Dr. Gerhard Bareiss, and to my parents Horst and Irene Schlaile. Last but not least, I am indebted to Carolin Becker and grateful for her sacrifices that helped me to finish this book.

I dedicate this book to my grandparents, who made so much possible: Hermann Bareiss (1916–2000), Ruth Bareiss (1921–2015), Rudolf Schlaile (1926–2009), and Hilde Schlaile (1924–2019).

Weissach im Tal, Germany Michael P. Schlaile
June 2020

Acknowledgment of previously published work

The editor, author(s), and publisher would like to acknowledge the following previously published material:

Chapter 4 was previously published with Elsevier (permission acquired via the Copyright Clearance Center's RightsLink service): Schlaile, M.P., Bogner, K., & Muelder, L. (2019). It's more than complicated! Using organizational memetics to capture the complexity of organizational culture. *Journal of Business Research.* https://doi.org/10.1016/j.jbusres.2019.09.035

Chapter 5 was previously published with Springer (open access, distributed under the terms of the Creative Commons Attribution 4.0 International License): Schlaile, M.P., Zeman, J., & Mueller, M. (2018). It's a match! Simulating compatibility-based learning in a network of networks. *Journal of Evolutionary Economics, 28* (5), 1111–1150. https://doi.org/10.1007/s00191-018-0579-z

Chapter 6 was previously published with Elsevier (open access, also distributed under the terms of the Creative Commons CC-BY license): Schlaile, M.P., Knausberg, T., Mueller, M., & Zeman, J. (2018). Viral ice buckets: A memetic perspective on the ALS Ice Bucket Challenge's diffusion. *Cognitive Systems Research, 52*, 947–969. https://doi.org/10.1016/j.cogsys.2018.09.012

Contents

1 **Editorial Introduction: We Are the Memes, Resistance Is Futile** ... 1
Michael P. Schlaile
   1.1 Positioning, Aims, and Scope 1
   1.2 Contribution and Structure of the Book 7
   References 9

2 **"Meme Wars": A Brief Overview of Memetics and Some Essential Context** 15
Michael P. Schlaile
   2.1 Episode I: The Phantom Memetics 15
   2.2 Episode II: Attack of the Meme Machines 18
   2.3 Episode III: Revenge of the Myth 19
   2.4 Episode IV: A New Replicator 20
   2.5 Episode V: The Interactor Strikes Back 22
   2.6 Episode VI: Return of the Memeplex 23
   References 24

3 **A Case for Econememetics? Why Evolutionary Economists Should Re-evaluate the (F)utility of Memetics** 33
Michael P. Schlaile
   3.1 Introduction 33
   3.2 Some Clarification and Classification 37
   3.3 Implications of the Informationalist Perspective on Memes 45
       3.3.1 No Meme is an Island: Memeplexes as Complex Systems 45
       3.3.2 The Economics of Attention: Memes in Competition for a Scarce Resource 47
   3.4 Rules and Imitation 48

	3.5	Conclusion: Building Bridges and Pointing the Way	51
		Appendix: Selected Examples of Definitions or Uses of Memes	53
		References	57
4	**It's More Than Complicated! Using Organizational Memetics to Capture the Complexity of Organizational Culture**		69
	Michael P. Schlaile, Kristina Bogner, and Laura Mülder		
	4.1	Introduction	70
	4.2	Memetics and Complexity	72
	4.3	Literature Review	73
	4.4	Case Example	75
		4.4.1 Sample and Method	75
		4.4.2 The Complex Memetic System of P3 Automotive	78
		4.4.3 Meme Map and Results	80
		4.4.4 Discussion and Limitations	84
	4.5	Conclusion and Avenues for Future Research	86
	Appendix		87
	References		94
5	**It's a Match! Simulating Compatibility-based Learning in a Network of Networks**		99
	Michael P. Schlaile, Johannes Zeman, and Matthias Mueller		
	5.1	Introduction	100
	5.2	Theoretical Background	102
		5.2.1 Relevant Literature and Research Focus	102
		5.2.2 Motivation and Foundations	104
		5.2.3 Model Description	106
	5.3	Results	110
		5.3.1 Baseline Analysis	110
		5.3.2 Effects of Knowledge Diversity	113
		5.3.3 Extended Analysis: Approaching the Qualitative Dimension of Knowledge	117
		5.3.4 The Importance of Being in the Right Place at the Right Time	119
		5.3.5 Results for Different Innovation Network Properties	122
	5.4	Conclusion and Outlook	126
	Appendix		129
	References		134

6 Viral Ice Buckets: A Memetic Perspective on the ALS Ice Bucket Challenge's Diffusion 141
Michael P. Schlaile, Theresa Knausberg, Matthias Mueller, and Johannes Zeman
- 6.1 Introduction 142
- 6.2 Theoretical Background 143
  - 6.2.1 Successful Memetic Replication and Diffusion 143
  - 6.2.2 The IBC as a Meme 146
- 6.3 Descriptive Memetic Analysis 148
  - 6.3.1 Origins and Examples of Variants of the IBC Meme 148
  - 6.3.2 Remarks on the IBC Meme's Diffusion Pattern 149
  - 6.3.3 Important Characteristics for the IBC's Diffusion 153
- 6.4 Agent-based Simulation of the IBC Diffusion 156
  - 6.4.1 Baseline Scenario 157
  - 6.4.2 Influence of Average Resistance 162
  - 6.4.3 Influence of Social Network Topology 164
  - 6.4.4 Influence of Celebrities 169
- 6.5 Discussion and Practical Implications 171
- 6.6 Conclusion and Outlook 174
- References 175

7 General Discussion: Economemetics and Agency, Creativity, and Normativity 181
Michael P. Schlaile
- 7.1 Agency—Memes as Quasi-Agents? 181
- 7.2 Everything is a Remix? A Note on Creativity and Novelty 185
- 7.3 Memethics: Normative Implications 189
- References 192

8 Conclusion and the Way(s) Forward 199
Michael P. Schlaile
- 8.1 General Limitations 199
- 8.2 Summary and Outlook 201
- References 203

Correction to: "Meme Wars": A Brief Overview of Memetics and Some Essential Context C1
Michael P. Schlaile

List of Figures

Fig. 3.1	The three-dimensional (P-I-E) perspective on memes.	42
Fig. 3.2	(Schematic) illustration of a generic complex population system of memes ("memeplex") with different sub-populations	46
Fig. 3.3	Institutions as the intersection of the sets of memes (M), habits (H), and rules (R).	51
Fig. 3.4	Illustration of compatibility relations between selected approaches and authors in evolutionary economics, memetics, and related fields	52
Fig. 4.1	Meme map of P3 automotive's organizational culture	82
Fig. 5.1	Illustration of the network-of-networks approach	107
Fig. 5.2	Illustration of the selection mechanism used in the random walk along the edges in a knowledge network B_i	109
Fig. 5.3	Average number of KUs $\langle N_B \rangle$ per agent over time in an innovation network A with an Erdős–Rényi (random network) topology for different compatibility thresholds γ	111
Fig. 5.4	Average number of KUs $\langle N_B \rangle$ per agent over time in an innovation network A with an Erdős–Rényi (random network) topology for different compatibility thresholds γ	112
Fig. 5.5	Distribution of compatibilities between pairs of KUs with $n_K = 32$ bits generated from an AKU with different bit reassignment probabilities p_k	115
Fig. 5.6	Average number of KUs $\langle N_B \rangle$ per agent over time in an innovation network A with an Erdős–Rényi (random network) topology for different KU bit reassignment probabilities p_k	115
Fig. 5.7	Average number of KUs $\langle N_B \rangle$ per agent over time in an innovation network A with an Erdős–Rényi (random network) topology for different AKU bit reassignment probabilities p_a	117

Fig. 5.8	Change of knowledge network properties over time in an innovation network A with an Erdős–Rényi (random network) topology for different AKU bit reassignment probabilities p_a.	119
Fig. 5.9	Change of individual knowledge network sizes N_{B_i} over time in an innovation network A with an Erdős–Rényi (random network) topology for a moderate level of cognitive distance ($p_a = 0.6$ and $p_k = 0.6$), knowledge exploitation by all agents and $\gamma = 0.75$	120
Fig. 5.10	Time series of relative deviations from the mean of different knowledge network properties depending on agents' harmonic closeness centrality H in the innovation network A for the case of a common knowledge background ($p_a = 0.0$) and a moderate level of knowledge heterogeneity ($p_k = 0.6$).	121
Fig. 5.11	Time series of relative deviations from the mean of different knowledge network properties depending on agents' harmonic closeness centrality H in the innovation network A for the case of highly diverse knowledge backgrounds ($p_a = 0.8$) and a moderate level of knowledge heterogeneity ($p_k = 0.6$).	122
Fig. 5.12	Time series of average knowledge network sizes $\langle N_B \rangle$ for different innovation network topologies. The different topologies are Barabási–Albert (BA), Erdős–Rényi (ER), and Watts–Strogatz (WS).	124
Fig. 5.13	Time series of average knowledge network sizes $\langle N_B \rangle$ for different innovation network topologies. The different topologies are Barabási–Albert (BA), Erdős–Rényi (ER), Watts–Strogatz (WS), and compatibility-based (CB)	126
Fig. 5.14	Distribution of compatibilities between pairs of randomly generated KUs for different numbers of bits n_K per KU	130
Fig. 5.15	Agents' number of KUs N_B with respect to their local properties in an innovation network A with Erdős–Rényi (random network) topology	132
Fig. 6.1	Replication loop according to Heylighen and Chielens (2009)	145
Fig. 6.2	Use of IBC-related hashtags on Twitter	150
Fig. 6.3	Worldwide Google search interest (popularity) for "ALS Ice Bucket Challenge"	151
Fig. 6.4	Google search interest (popularity) of "ALS ice bucket challenge" compared between Canada, USA, Germany, Russia, and the UK	152
Fig. 6.5	Daily donations to the ALS Association	153
Fig. 6.6	Visualization of the simulated nomination procedure.	159
Fig. 6.7	Time series for BA networks	160

List of Figures

Fig. 6.8	Histogram for the number of carriers who accepted the challenge at the end of the simulation in a BA-type network..	161
Fig. 6.9	Newly accepted challenges over time for different values of mean resistance in BA-type networks with 500 agents and average degree of 150.............................	163
Fig. 6.10	Left panel: Final number of accepted challenges at the end of the simulation as a function of mean resistance; right panel: critical mass needed for the challenge to reach all agents in the network as a function of mean resistance (in a BA-type network)..	163
Fig. 6.11	Comparison of diffusion dynamics on three different network types...	165
Fig. 6.12	Comparison of total number of accepted challenges at the end of the simulation as a function of average degree for three different network types.................................	166
Fig. 6.13	Comparison of average network indices as a function of average degree for three different network types...........	167
Fig. 6.14	Comparison of diffusion dynamics with and without celebrities..	170

List of Tables

Table 4.1	Meme frequency (#) and (min-max-normalized) SBS of individual memes	81
Table 4.2	Key contributions to the field of organizational memetics	88
Table 5.1	Several global properties of the examined innovation networks A	124
Table 6.1	Selected properties of the networks	164

Chapter 1
Editorial Introduction: We Are the Memes, Resistance Is Futile

Michael P. Schlaile

Abstract This chapter provides a general editorial introduction to the book, positions it in the literature, and explains its rationale and contribution. This book is a collection of previously unpublished and already published work on what the author calls "econememetics". All in all, this book presents a theoretical foundation and selected cases for econememetics, which is argued to have the potential to build bridges between various disciplines and approaches and is, therefore, suitable for combating the "siloism" or fragmentation of evolutionary economics.

1.1 Positioning, Aims, and Scope

This book is about a particular perspective on cultural and economic evolution called *meme theory* or *memetics*.[1] Whether or not that perspective is worthy of serious academic investigation has been quite controversial. To get straight to the point: I (still) think it is.[2] Before we can delve into this highly interdisciplinary topic, a few introductory remarks are called for that attempt to position this book in the extensive literature and delineate the aims and scope of the exploration that follows.

---

[1] For the purpose of a working definition, in this introductory section we take up the *Oxford Dictionary of English*, where the meme is (primarily) defined as "an element of a culture or system of behaviour passed from one individual to another by imitation or other non-genetic means" (https://en.oxforddictionaries.com/definition/meme; see also Stevenson 2010). For the term memetics, we follow the *Merriam-Webster* dictionary, which defines it simply as "the study of memes" (https://www.merriam-webster.com/dictionary/memetics).

[2] Others have quite recently argued along the same lines (e.g., Boudry 2018a, b; Boudry and Hofhuis 2018; Dennett 2017, 2018; Hofhuis and Boudry 2019; Stewart-Williams 2018; van den Bergh 2018), and I wholeheartedly agree with Steije Hofhuis and Maarten Boudry that a "wholesale dismissal of memetics may be premature" (Hofhuis and Boudry 2019, p. 14).

M. P. Schlaile (✉)
Department of Innovation Economics, University of Hohenheim, Stuttgart, Germany
e-mail: schlaile@uni-hohenheim.de

To understand the motivation for this book, we first have to take a step back and remember that culture and economic processes, agents, and organizations are highly intertwined (e.g., Beugelsdijk and Maseland 2011; Granovetter 2017; Herrmann-Pillath 2000, 2010; Weber 1930; Zelizer 2011). As Michael Schramm puts it, culture can be regarded as a software[3] for the economic hardware or the "identity semantics" of economic agents (cf. Schramm 2008, pp. 13, 61). More precisely, according to Schramm (2008), cultures are necessarily "economic moral cultures" because cultural values clearly affect what is considered to be right/wrong, good/bad, or proper/improper behavior (in short: *morality*), while these moral cultures are continuously subjected to economic pressures and must "prove themselves" *economically*. Investigations into cultural influences on business and the economy abound, and the field has advanced both empirically and conceptually over the past decades (e.g., Baumann Montecinos 2019; Hofstede 2001; Hofstede et al. 2010; House et al. 2004; Inglehart 2018; Trompenaars and Hampden-Turner 2012).[4] My own interest in this broad and interdisciplinary topic was sparked almost 10 years ago especially by the lectures and works of Michael Schramm (e.g., 2008) and Eugen Buß (e.g., 1996, 2012), prompting me to look closer into the influence of culture on (global) leadership (Schlaile 2012). After all, as Amartya Sen writes: "The real issue … is *how*—not whether—culture matters" (Sen 2004, p. 38, italics in original).

Another piece in the puzzle is the dynamic and *evolutionary* character of both culture and the economy.[5] My interest in evolutionary approaches to culture was piqued around the second half of 2012 more or less coincidentally by stumbling upon the meme concept, which was initially promoted in *The Selfish Gene* by Richard Dawkins (1976, 2016) and quite prominently refined by Susan Blackmore (1999) in *The Meme Machine*, by Daniel Dennett (1995) in *Darwin's Dangerous Idea*, and by a number of other so-called "memeticists" elsewhere (e.g., Aunger 2000, 2002; Blute 2010; Brodie 1996; Cullen 2000; Distin 2005; Lynch 1996; Tyler 2011; or in the *Journal of Memetics*, which was published from 1997 to 2005).[6] Simultaneously, I

---

[3] Perhaps "operating system" could be a more appropriate metaphor, but Schramm seems to follow Hofstede et al. (2010) here, who popularized the "software of the mind" analogy with the title of their well-known and frequently cited book. Interestingly, Runciman (2009, p. 54) argues in the same vein that a "plausible way of looking at memes is to see them as the software of the hard-wired human brain."

[4] See also https://geerthofstede.com, https://globeproject.com, http://www.worldvaluessurvey.org, for an extensive stock of resources and references.

[5] Note that *evolution(ary)* can mean rather different things ranging from synonyms to change and development to more nuanced connotations in terms of biological analogies and concepts. As Geoffrey Hodgson puts it, "the word 'evolutionary' is extremely vague. It is now widely used, even by economists using neoclassical techniques … Above all, 'evolutionary' is now a voguish word that everyone seems keen to use. In precise terms it signifies little or nothing" (Hodgson 2000, p. 322). In this regard, there seems to be no real consensus even among evolutionary economists (e.g., see Callebaut 2011; Dopfer 2001; and Potts 2009; Hodgson 1993; Hodgson and Lamberg 2016; Nelson and Winter 1982; Nelson et al. 2018; Witt 2003, 2014, 2016). Therefore, the term evolution(ary) will be kept deliberately vague in this introductory chapter, but it shall become clear that, over the course of this book, it gets a particular (neo-)Darwinian connotation.

[6] See http://cfpm.org/jom-emit/ for an archive of all articles published in that journal.

was lucky to have been introduced to the world of (neo-Schumpeterian) evolutionary economics by Andreas Pyka (e.g., Hanusch and Pyka 2007; Pyka 2019; Pyka and Hanusch 2006) and other colleagues.

By connecting the dots and bringing together the insights that culture matters for business and the economy and that evolutionary processes (especially involving *variation, selection,* and *retention*)[7] govern the dynamic interplay, the "real issue" of *how* culture matters can arguably be addressed.[8] Hence, this book joins the ranks of numerous works on cultural and economic evolution. In a way, many of the topics discussed here can already be traced back to prominent writers of the late nineteenth and early twentieth century such as Joseph Schumpeter (e.g., 1912/2006), Gabriel Tarde (e.g., 1903), Thorstein Veblen (e.g., 1898, 1899), Max Weber (e.g., 1930, 1922/2013),[9] and many others.

The insight that cultural values, ideas, knowledge, habits, and so on play a role on many levels of society and economy is neither new nor particularly controversial. Debates and points of contention quite naturally arise, however, when it comes to the age-old question of the social sciences whether it is mainly the social structure or human agency that determines behavior. Indeed, this brings us to the first way of positioning this book within the spectrum of approaches and perspectives. Christian Illies (2005, 2010) has distinguished five types of evolutionary explanations with regard to culture, which we can use for a first classification:[10]

- *Natural boundary conditions for cultural evolution (Type A).* Here, the focus is on environmental conditions (e.g., climate) as well as the biological nature of humans in the sense of a frame for actualizing culture. In the latter regard, Illies gives a pithy example: If Homo sapiens had not evolved legs, there would also be no waltz. In short, Type A defines general boundaries and the opportunity space for culture and its evolution.
- *Enabling predispositions for human culture (Type B).* This type frequently falls within the scope of sociobiology and evolutionary psychology and focuses on behavioral dispositions for (usually adaptive, i.e., genetic fitness enhancing) culture in terms of rules, strategies, social norms, etc. Illies mentions examples like dispositions for cooperation and kin selection, but also nepotism, inheritance laws, incest taboos, etc. He quotes Edward O. Wilson for clarification: "Genes prescribe epigenetic rules, which are the regularities of sensory perception and mental development that animate and channel the acquisition of culture" (Wilson 1998, p. 171, as cited in Illies 2010, p. 20).

---

[7]This Darwinian triad is at the heart of *generalized Darwinism* (e.g., Aldrich et al. 2008; see also Campbell 1965); note that instead of "retention" one may also find terms such as inheritance, heredity, reproduction, or transmission.

[8]One should condone the overuse of Sen's (2004) quote beyond its original context, here.

[9]As, for example, Calhoun et al. (2012, p. 267) explain: "Weber was also profoundly interested in the cultural orientations of social actors. For Weber, ideas and value orientations ... were important because they motivate action. Although ideas may be shaped by material conditions, Weber held that the reverse might also be true."

[10]Note, however, that Illies himself cautions against viewing these types as being clearly demarcated or unambiguous (cf. Illies 2010, pp. 19, 27).

- *Cultural evolution as a semi-autonomous evolutionary process (Type C).* Here, the emphasis is placed on the feedback between the evolutionary success of a culture and the survival of the respective group and its members. Illies lists examples such as fighting rituals, superior customs and traditions, and (military) strategies that may in turn influence the propagation of (the genes of) a particular group. Put differently, Type 3 essentially comprises so-called *dual inheritance theory* and *gene-culture coevolution* (e.g., see Durham 1991; Laland and Brown 2011; Newson and Richerson 2018, for overviews).[11]
- *Cultural evolution as an autonomous evolutionary process (Type D).* According to Illies, an autonomy of cultural evolution is given when there is no positive influence of a cultural evolutionary process on the biological fitness of the culture's biological carrier(s). Among others, Illies mentions *memetics* (referring to Blackmore 1999, and Dawkins 1976) as an example of a Type D approach and explains that cultural evolution can be not just independent from the reproductive success of that culture's human carriers but even seriously maladaptive (e.g., as in the case of sexual abstinence norms). In short, Type D can be said to focus on processes of selection and competition that lead to the reproduction or replication of particular cultural elements without them having a positive influence on the reproductive success of the people practicing or "carrying" these cultural elements.[12]
- *Evolutionary theory as an interpretation of the world (Type E).* According to Illies, some scholars advocate evolutionary theory as a generalizable and universal worldview (see also Buskes 2008) in the sense of a "theory of everything" (Illies 2010, p. 25). He mentions authors such as Dennett (1995) and E.O. Wilson (1998); and we might as well add other proponents of an evolutionary social science (e.g., Blute 2010; Turner et al., 2015; D.S. Wilson 2019b), *generalized Darwinism* (e.g., Hodgson and Knudsen 2010), or even "generalized evolutionism" (e.g., Tang 2017). As Illies claims, Type E differs from Types A–D inasmuch as it does not aim at explaining how specific mechanisms lead to the evolution of a particular cultural phenomenon. Instead, it is argued that one central cultural phenomenon, our interpretation of the world, is well-founded in evolutionary theory.[13]

This book can be *mainly* positioned within Type D,[14] while also taking on an overall evolutionary Type E perspective. That being said, I do not want to create the impression that I consider a Type D perspective to be the only correct one, or, for

---

[11] In this regard, it seems appropriate to complete the above quote by E.O. Wilson (1998, p. 171): "Culture helps to determine which of the prescribing genes survive and multiply from one generation to the next. Successful new genes alter the epigenetic rules of populations. The altered epigenetic rules change the direction and effectiveness of the channels of cultural acquisition."

[12] As Paolo Inghilleri (1999, p. 7) puts it, in this case "there is not a regulatory hierarchical bond going from the biological to the cultural. We are now faced with two distinct systems of evolution of information: biological evolution and cultural evolution."

[13] Thereby, according to Illies (2010), Type E can be conceived as either building on Types A–D or an overarching perspective, which clearly relates to *evolutionary epistemology*.

[14] "Mainly" meaning that cultural evolution will be regarded as an independent evolutionary process in the sense that it does not necessarily serve the actors' *economic* survival, utility, or interests. Yet, we will not be concerned with the genes of the economic agents in this book.

that matter, that memetics holds the answer to every question.[15] Quite the contrary, as we know from the literature (e.g., Henrich 2016) and our daily experiences, many culturally evolved practices, habits, institutions, technological innovations, etc., are certainly helpful for us as individuals, as groups, maybe even as a species (and often also for our genes). Yet, I side here with those who bring up the painful but important subject that evolution does not always mean *progress* (e.g., Kappelhoff 2012; Mayr 2001; Ruse 1993; Wuketits 1997), and that innovations are not *per se* good for us (incidentally, this is a recurring theme in the *responsible innovation* literature; see Schlaile et al., 2018, and references therein). Moreover, one does not even have to be an adherent of memetics to acknowledge that cultural ideas can have a quite independent existence. As, for example, John Kenneth Galbraith explains with regard to *conventional wisdom*:

> The first requirement for an understanding of contemporary economic and social life is a clear view of the relation between events and the ideas which interpret them. For each of these has *a life of its own* and ... each is capable for a considerable period of *pursuing an independent course* (Galbraith 1998, p. 6, emphasis added).

Likewise, as David Sloan Wilson (2002, p. 53) reminds us, Émile Durkheim already "anticipates the modern concept of 'selfish memes'",[16] which may be illustrated by the following quote:

> The totality of beliefs and sentiments common to the average members of a society forms a determinate system *with a life of its own* (Durkheim (2013) [1893], p. 63, emphasis added).

Alternatively, in the words of Erich Kahler:

> Cultures are not identical with their originating peoples and historical spheres; they are their offspring, their *spiritual spores* as it were. They mature very late and come into being as *detached, separate units* of history only in the ultimate stages of their originators. In this capacity, as *independent entities,* intermediate between ethnic communities and man, they represent and carry evolution (Kahler 1968, p. 18, emphasis added).

And perhaps one of the most daunting views in this regard was presented by Ted Cloak:

> In a human carrier ... a cultural instruction is more analogous to a viral or bacterial gene than to a gene of the carrier's own genome. It is like an active parasite that controls some behavior of its host. ... In short, 'our' cultural instructions don't work for us organisms; we work for them. At best, we are in symbiosis with them, as we are with our genes. At worst, we are their slaves (Cloak 1975, p. 172).

---

[15] Arguably, it is sometimes better to have different perspectives on a particular issue than just one all-encompassing theory; think of the parable of the blind men and the elephant. As also David Sloan Wilson (2015, Chap. 3) explains using the metaphors of accounting methods, viewing a mountain from different vantage points, and speaking different languages, divergent configurations of ideas deserve to coexist and need not necessarily be incommensurable—although they can be. The notion of *equivalence*, which D.S. Wilson (2015, 2019a) uses, may not be fully appropriate in the above context of Types A–E mentioned by Illies (2010) because the respective theories and approaches differ not only in terms of perspectives but sometimes also by invoking different causal mechanisms. Yet, I would argue that all have their merits and limits and their right to coexist—and potentially complement each other.

[16] See also Distin (1995) on a related argument.

Finally, to take up a more recent quote from Yuval Noah Harari:

> No matter what you call it – game theory, postmodernism or memetics – the dynamics of history are not directed towards enhancing human well-being. There is no basis for thinking that the most successful cultures in history are necessarily the best ones for *Homo sapiens*. Like evolution, history disregards the happiness of individual organisms (Harari 2015, p. 271, italics in original).

This list could go on, but at this point it should suffice for making the point that there are sometimes good reasons for adopting a Type D perspective on cultural evolution as an autonomous evolutionary process. What does this perspective mean for economics? Indeed, it is precisely that question which motivated me to write this book, which is a revised and extended version of my (cumulative, paper-based) doctoral dissertation. It should be noted, though, that the concept of memes has actually been used before to explicitly address economic questions and issues (e.g., Breitenstein 2002). I must admit, therefore, that the (sub)title of this book may be slightly misleading in that regard. Having said that, to the best of my knowledge, this book is the first targeted attempt to investigate what *evolutionary* economics may gain from a memetic perspective (and probably vice versa). More precisely, the overarching research question of this book and all of its chapters is

> (How) can memetics be fruitfully utilized for evolutionary economics?

Since this question would be broad enough to fuel several research projects for many years, I have decided to present the current state of (my) research into this highly dynamic topic by limiting myself to the following four selected issues or sub-questions within this book:

1. The first study is essentially motivated by the fragmentation of evolutionary economics as a discipline. I have, therefore, asked myself *Q1: (How) can a memetic approach to economic evolution help to reveal links and build bridges between different but complementary concepts and approaches in evolutionary economics and related disciplines?*
2. The second study is motivated by a hitherto unexploited potential of using a memetic approach to capturing the complexity of organizational culture. *Q2: (How) can the diversity and interdependence of characteristic elements of an organizational culture be captured such that these elements can be considered as representations of the underlying organizational memes?*
3. The third study brings a memetic perspective to a central topic within evolutionary economics, namely, the diffusion of knowledge in innovation systems. *Q3: How can we model the diffusion of knowledge in innovation networks while taking a memetic representation of knowledge seriously?*
4. The fourth and final question addressed in this book is essentially twofold and motivated by the need for a study of an actual diffusion phenomenon from a memetic perspective. The 2014 "viral" Internet phenomenon called *ALS Ice Bucket Challenge* serves as an appropriate case study to address the following interrelated questions. $Q4_a$: *Which endogenous (meme-centered) elements of the Ice Bucket Challenge meme contributed to its diffusion pattern?* and $Q4_b$: *Can we*

identify structural factors (i.e., network properties) that influenced the diffusion of the Ice Bucket Challenge meme on social networks?

In an early attempt to articulate my initial take on memes as optimizing entities maximizing their own "utility" (Schlaile 2013),[17] I came up with the term *econometics* to denote this interdisciplinary endeavor of synergizing memetics and (evolutionary) economics. Only that I did not really coin that term, as I found out soon after, because Aaron Lynch (2000) already wrote in a paper about two decades ago: "The study of how economic and memetic forces combine and interact is a broad interdisciplinary field I call econo-memetics" (Lynch 2000, pp. 14–15). Apparently, however, econometics was not pursued much further after Lynch coined the field,[18] so that this book may well be one of the first steps in that direction.[19] In the following sub-section, the structure and central contributions of this book are summarized with the intention of giving a teaser for the chapters that follow.

1.2 Contribution and Structure of the Book

First and foremost, the chapters of this volume contribute to the literature on econometics in the above-mentioned sense. Who might be interested in reading this book, and what will we have gained? By the end of Chap. 8, I will have achieved two things: (1) I have made the case that memetics still bears a serious conceptual and empirical research potential (not only) for evolutionary economics, and (2) I have shown with the help of my co-authors how network science can be used to analyze memes and their diffusion in a population of "carriers."

It should be quite obvious by now that this book makes a primarily theoretical contribution to an interdisciplinary field and mainly targets an academic audience of evolutionary, institutional, cultural, and complexity economists as well as scholars interested in cultural evolution, transmission, and diffusion. In other words, this book presents ideas and approaches that should be read as an invitation for further interdisciplinary work in this direction. Nevertheless, for practitioners, the insights

---

[17]For example, in Schlaile (2013), I proposed that a certain meme m_i's utility $u(m_i)$ could be interpreted as a measure of its success in replication, which is primarily depending on its longevity η_i, fecundity ξ_i, and copying-fidelity γ_i: $u(m_i) = \eta_i \cdot \xi_i \cdot \gamma_i$, with η_i ($0 < \eta_i$), where 0 means "never existed," with ξ_i ($0 < \xi_i$) as the rate of replication, where 0 means "infertile," and with γ_i ($0 < \gamma_i \leq 1$), where 0 means "copying impossible" and 1 denotes a "flawless copy."

[18]Which is probably also due to Lynch's untimely passing in 2005.

[19]Note, however, that Tony Yates (2019) claims to have written a (hitherto unpublished) paper on memetics and economics already in the early 2000s, and Rolf Breitenstein's (2002) German book on *Memetics and Economics: How Memes Determine Markets and Organizations* can definitely be counted as the first book on econometics. Yet, Breitenstein's (2002) work has a much broader conceptual orientation than the present book, and he does not explicitly target the evolutionary economics literature nor attempts to establish serious connections with it (indeed, he seems to almost brush aside the whole field with some brief remarks and pointers to Esben Sloth Andersen, Kenneth Boulding, Geoffrey Hodgson, Stan Metcalfe, Paolo Saviotti, and Ulrich Witt in a couple of footnotes on page 62).

gained especially from the chapters on organizational culture and the diffusion of the Ice Bucket Challenge may be applied to similar managerial issues (e.g., how to measure and potentially facilitate the propagation of desirable cultural information and impede the dissemination of unwanted content). For policy-makers, the discussion on the (lack of) optimal innovation network structures in the face of different degrees of knowledge diversity indicates that innovation policy should put more emphasis on adequate *knowledge* policies.[20]

In the end, this book probably raises even more questions—in terms of suggestions for future research—than it actually answers; but identifying promising next steps and proposing links to related fields and approaches is arguably an important contribution to the advancement of a scientific field in itself. That said, the book makes specific contributions in order to answer the above-mentioned (sub-)research questions. The common theme connecting these rather diverse contributions is a memetic perspective.

The following Chap. 2 sets the stage for the subsequent treatises by providing some essential context and a brief overview over the history and current issues in memetics. Given the book's focus, this overview does not claim to be exhaustive.

Chapter 3 approaches *Q1*. The chapter's main contribution is that it reveals and establishes connections between various rather fragmented lines of research. Motivated by the starting point that both imitation and cultural evolution have not received sufficient attention from evolutionary economists, I review and clarify some points of criticism (Sect. 3.2) and propose to follow an informationalist approach (Sect. 3.3) to memes that is also in line with the notion of complex population systems. Moreover, I am able to establish links to imitation heuristics, evolutionary institutionalism, and the rule-based approach (Sect. 3.4).

Chapter 4 tackles *Q2*. This chapter contributes to bridging the gap between theorizing and empirical research on so-called "organizational memetics" by presenting a state-of-the-art review of the literature on organizational memetics (Sect. 4.3) and demonstrating how meme maps can be used to highlight interdependencies among organizational memes (Sect. 4.4).

Chapter 5 deals with *Q3*. The chapter primarily contributes to the literature on knowledge diffusion and assimilation in innovation networks by developing a new agent-based simulation model that is designed to incorporate three essential characteristics of knowledge also known from the memetics literature: that knowledge can be regarded as a network of knowledge units or memes, that new memes are more likely to be adopted when they are compatible with already existing ones, and that their transmission requires attention. More precisely, the chapter reviews important literature, elucidates the research focus, motivation, and foundations, and then describes the agent-based simulation model in Sect. 5.2. The results of the simulations are presented and discussed in Sect. 5.3.

---

[20]What more adequate knowledge policies might imply is discussed in the context of a sustainable knowledge-based bioeconomy in an article I co-authored with my colleagues Sophie Urmetzer, Kristina Bogner, Matthias Müller, and Andreas Pyka (2018).

Chapter 6 addresses *Q4* and contributes to research on social learning and social contagion by presenting a case study on the diffusion of the Ice Bucket Challenge, thereby narrowing the gap between meme-centered perspectives and social network analysis. In Sect. 6.2, the necessary theoretical background and the suitability of a memetic perspective are discussed. Section 6.3 presents a descriptive memetic analysis, whereas the second pillar of the chapter contributes to the literature by developing and using an agent-based simulation model (Sect. 6.4) to analyze different scenarios and parameter settings. In Sect. 6.5, the findings, limitations, and practical implications of the chapter are discussed.

In Chap. 7, a general discussion contributes to tying up a few loose ends about "selfish memes" and agency (Sect. 7.1), creativity (Sect. 7.2), and normativity (Sect. 7.3).

Subsequently, the book concludes with Chap. 8 by addressing general limitations in Sect. 8.1 and by summarizing the key findings and proposing avenues for future research in Sect. 8.2.

Note that, due to the simple fact that Chaps. 3–6 have been developed as independent articles, and three of them have been previously published, some repetition and slight variations in writing style should be expected.

References

Aldrich, H. E., Hodgson, G. M., Hull, D. L., Knudsen, T., Mokyr, J., & Vanberg, V. J. (2008). In defence of generalized Darwinism. *Journal of Evolutionary Economics*, *18*, 577–596. https://doi.org/10.1007/s00191-008-0110-z.

Aunger, R. (Ed.). (2000). *Darwinizing culture: The status of memetics as a science*. Oxford: Oxford University Press.

Aunger, R. (2002). *The electric meme: A new theory about how we think*. New York: Free Press.

Baumann Montecinos, J. (2019). *Moralkapital und wirtschaftliche Performance: Informelle Institutionen, Kooperation, Transkulturalität*. Wiesbaden: Springer.

Beugelsdijk, S., & Maseland, R. (2011). *Culture in economics: History, methodological reflections, and contemporary applications*. Cambridge: Cambridge University Press.

Blackmore, S. (1999). *The meme machine*. Oxford: Oxford University Press.

Blute, M. (2010). *Darwinian sociocultural evolution: Solutions to dilemmas in cultural and social theory*. Cambridge: Cambridge University Press.

Boudry, M. (2018a). Invasion of the mind snatchers. On memes and cultural parasites. *Teorema*, *37*(2), 111–124.

Boudry, M. (2018b). Replicate after reading: On the extraction and evocation of cultural information. *Biology & Philosophy*, *33*. https://doi.org/10.1007/s10539-018-9637-z.

Boudry, M., & Hofhuis, S. (2018). Parasites of the mind. Why cultural theorists need the meme's eye view. *Cognitive Systems Research*, *52*, 155–167. https://doi.org/10.1016/j.cogsys.2018.06.010.

Breitenstein, R. (2002). *Memetik und Ökonomie: Wie die Meme Märkte und Organisationen bestimmen*. Münster, Germany: Lit.

Brodie, R. (1996). *Virus of the mind. The new science of the meme*. Seattle: Integral Press.

Buskes, C. (2008). *Evolutionär denken: Darwins Einfluss auf unser Weltbild*. Darmstadt: Primus Verlag.

Buß, E. (1996). *Lehrbuch der Wirtschaftssoziologie* (2nd ed.). Berlin: de Gruyter.

Buß, E. (2012). *Managementsoziologie: Grundlagen, Praxiskonzepte, Fallstudien* (3rd ed.). München: Oldenbourg.
Calhoun, C. J., Gerteis, J., Moody, J., Pfaff, S., & Virk, I. (Eds.). (2012). *Classical sociological theory* (3rd ed.). Chichester, West Sussex: Wiley.
Callebaut, W. (2011). Beyond generalized Darwinism. I. Evolutionary economics from the perspective of naturalistic philosophy of biology. *Biological Theory, 6*(4), 338–350. https://doi.org/10.1007/s13752-013-0086-2.
Campbell, D. T. (1965). Variation and selective retention in socio-cultural evolution. In H. R. Barringer, G. I. Blanksten, & R. W. Mack (Eds.), *Social change in developing areas: A reinterpretation of evolutionary theory* (pp. 19–49). Cambridge, MA: Schenkman.
Cloak, F. T. (1975). Is a cultural ethology possible? *Human Ecology, 3*(3), 161–182. https://doi.org/10.1007/BF01531639.
Cullen, B. (2000). *Contagious ideas: On evolution, culture, archaeology, and cultural virus theory. Collected writings* (J. Steele, R. Cullen, & C. Chippindale (Eds.)). Oxford: Oxbow Books.
Dawkins, R. (1976). *The selfish gene*. Oxford: Oxford University Press.
Dawkins, R. (2016). *The selfish gene* (40th anniversary ed.). Oxford: Oxford University Press.
Dennett, D. C. (1995). *Darwin's dangerous idea: Evolution and the meanings of life*. New York: Simon & Schuster.
Dennett, D. C. (2017). *From bacteria to Bach and back: The evolution of minds*. New York: W.W. Norton & Company.
Dennett, D. C. (2018). Comment on Boudry. *Teorema, 37*(3), 125–127.
Distin, K. (1995). Durkheim: social facts as memes? (unpublished manuscript).
Distin, K. (2005). *The selfish meme. A critical reassessment*. Cambridge: Cambridge University Press.
Dopfer, K. (2001). Evolutionary economics - framework for analysis. In K. Dopfer (Ed.), *Evolutionary economics: Program and scope* (pp. 1–44). New York: Springer.
Dopfer, K., & Potts, J. (2009). On the theory of economic evolution. *Evolutionary and Institutional Economics Review, 6*(1), 23–44. https://doi.org/10.14441/eier.6.23.
Durham, W. H. (1991). *Coevolution: Genes, culture, and human diversity*. Stanford: Stanford University Press.
Durkheim, É. (2013). *The division of labour in society* (2nd ed.). Edited and with a new introduction by Steven Lukes. Translation by W.D. Halls. Basingstoke: Palgrave Macmillan.
Galbraith, J. K. (1998). *The affluent society* (40th anniversary ed.). Updated and with a new introduction by the author. Boston: Mariner Books.
Granovetter, M. (2017). *Society and economy: Framework and principles*. Cambridge, MA: Harvard University Press.
Hanusch, H., & Pyka, A. (Eds.). (2007). *Elgar companion to neo-Schumpeterian economics*. Cheltenham (UK): Edward Elgar.
Harari, Y. N. (2015). *Sapiens: A brief history of humankind*. London: Vintage Books.
Henrich, J. (2016). *The secret of our success: How culture is driving human evolution, domesticating our species and making us smarter*. Princeton: Princeton University Press.
Herrmann-Pillath, C. (2000). *Evolution von Wirtschaft und Kultur: Bausteine einer transdisziplinären Methode*. Marburg: Metropolis.
Herrmann-Pillath, C. (2010). What have we learnt from 20 years of economic research into culture? *International Journal of Cultural Studies, 13*(4), 317–335. https://doi.org/10.1177/1367877910369966.
Hodgson, G. M. (1993). *Economics and evolution: Bringing life back into economics*. Cambridge: Polity Press.
Hodgson, G. M. (2000). What is the essence of institutional economics? *Journal of Economic Issues, 34*(2), 317–329. https://doi.org/10.1080/00213624.2000.11506269.
Hodgson, G. M., & Knudsen, T. (2010). *Darwin's conjecture: The search for general principles of social and economic evolution*. Chicago: University of Chicago Press.

Hodgson, G. M., & Lamberg, J.-A. (2018). The past and future of evolutionary economics: Some reflections based on new bibliometric evidence. *Evolutionary and Institutional Economics Review, 15*(1), 167–187. https://doi.org/10.1007/s40844-016-0044-3.

Hofhuis, S., & Boudry, M. (2019). 'Viral' hunts? A cultural Darwinian analysis of witch persecutions. *Cultural Science Journal, 11*(1), 13–29. https://doi.org/10.5334/csci.116.

Hofstede, G. (2001). *Culture's consequences: Comparing values, behaviors, institutions, and organizations across nations* (2nd ed.). Thousand Oaks: Sage.

Hofstede, G., Hofstede, G. J., & Minkov, M. (2010). *Cultures and organizations: Software of the mind. Intercultural cooperation and its importance for survival* (3rd ed.). New York: McGraw Hill.

House, R. J., Hanges, P. J., Javidan, M., Dorfman, P. W., & Gupta, V. (Eds.). (2004). *Culture, leadership, and organizations: The GLOBE study of 62 societies.* Thousand Oaks, CA: Sage.

Illies, C. (2005). Die Gene, die Meme und wir. In B. Goebel, A. M. Hauk, & G. Kruip (Eds.), *Probleme des Naturalismus* (pp. 127–160). Paderborn: Mentis.

Illies, C. (2010). Biologie statt Philosophie? Evolutionäre Kulturerklärungen und ihre Grenzen. In V. Gerhardt & J. Nida-Rümelin (Eds.), *Evolution in Natur und Kultur* (pp. 15–38). Berlin: de Gruyter.

Inghilleri, P. (1999). *From subjective experience to cultural change.* transl. by E. Bartoli, originally published as Esperienza soggettiva, personalità, evoluzione culturale. Cambridge: Cambridge University Press.

Inglehart, R. (2018). *Cultural evolution: People's motivations are changing, and reshaping the world.* Cambridge: Cambridge University Press.

Kahler, E. (1968). Culture and evolution. In M. F. A. Montagu (Ed.), *Culture: Man's adaptive dimension* (pp. 3–19). New York: Oxford University Press.

Kappelhoff, P. (2012). Selektionsmodi der Organisationsgesellschaft: Gruppenselektion und Memselektion. In S. Duschek, M. Gaitanides, W. Matiaske, & G. Ortmann (Eds.), *Organisationen regeln: Die Wirkmacht korporativer Akteure* (pp. 131–162). Wiesbaden: Springer.

Laland, K. N., & Brown, G. R. (2011). *Sense and nonsense: Evolutionary perspectives on human behaviour* (2nd ed.). Oxford: Oxford University Press.

Lynch, A. (1996). *Thought contagion: How belief spreads through society. The new science of memes.* New York: Basic Books.

Lynch, A. (2000). Thought contagions in the stock market. *Journal of Psychology and Financial Markets, 1*(1), 10–23. https://doi.org/10.1207/S15327760JPFM0101_03.

Mayr, E. (2001). *What evolution is.* New York: Basic Books.

Nelson, R. R., Dosi, G., Helfat, C., Pyka, A., Saviotti, P. P., Lee, K., et al. (2018). *Modern evolutionary economics: An overview.* Cambridge: Cambridge University Press.

Nelson, R. R., & Winter, S. G. (1982). *An evolutionary theory of economic change.* Cambridge: The Belknap Press of Harvard University Press.

Newson, L., & Richerson, P. (2018). Dual inheritance theory. In H. Callan (Ed.), *The international encyclopedia of anthropology* (pp. 1–5). https://doi.org/10.1002/9781118924396.wbiea1881

Pyka, A. (2019). Evolutorische Innovationsökonomik. In B. Blättel-Mink, I. Schulz-Schaeffer, & A. Windeler (Eds.), *Handbuch Innovationsforschung* (pp. 1–19). https://doi.org/10.1007/978-3-658-17671-6_6-1

Pyka, A., & Hanusch, H. (2006). Introduction. In A. Pyka & H. Hanusch (Eds.), *Applied evolutionary economics and the knowledge-based economy* (pp. 1–9). Cheltenham: Edward Elgar.

Runciman, W. G. (2009). *The theory of cultural and social selection.* Cambridge: Cambridge University Press.

Ruse, M. (1993). Evolution and progress. *Trends in Ecology & Evolution, 8*(2), 55–59. https://doi.org/10.1016/0169-5347(93)90159-M.

Schlaile, M.P. (2012). Global Leadership im Kontext ökonomischer Moralkulturen - eine induktiv-komparative Analyse. *Hohenheimer Working Papers zur Wirtschafts- und Unternehmensethik* Nr. 13. Retrieved from https://theology-ethics.uni-hohenheim.de/fileadmin/einrichtungen/theology-ethics/hhwpwue_13_Schlaile.pdf

Schlaile, M.P. (2013). *A 'more evolutionary' approach to economics: The Homo sapiens oeconomicus and the utility maximizing meme*. Paper presented at the 11th Globelics International Conference on entrepreneurship, innovation policy and development in an era of increased globalisation, September 11–13. Middle East Technical University, Ankara, Turkey.

Schlaile, M. P., Mueller, M., Schramm, M., & Pyka, A. (2018). Evolutionary economics, responsible innovation and demand: Making a case for the role of consumers. *Philosophy of Management, 17*(1), 7–39. https://doi.org/10.1007/s40926-017-0054-1.

Schramm, M. (2008). *Ökonomische Moralkulturen. Die Ethik differenter Interessen und der plurale Kapitalismus*. Marburg: Metropolis.

Schumpeter, J. A. (1912/2006). *Theorie der wirtschaftlichen Entwicklung. Nachdruck der 1. Auflage von 1912*. Herausgegeben und ergänzt um eine Einführung von Jochen Röpke und Olaf Stiller. Berlin: Duncker & Humblot.

Sen, A. (2004). How does culture matter? In V. K. Rao & M. Walton (Eds.), *Culture and public action*. Stanford, CA: Stanford University Press.

Stevenson, A. (Ed.). (2010). *The Oxford dictionary of English* (3rd ed.). Oxford: Oxford University Press.

Stewart-Williams, S. (2018). *The ape that understood the universe: How the mind and culture evolve*. Cambridge: Cambridge University Press.

Tang, S. (2017). Toward generalized evolutionism: Beyond "generalized Darwinism" and its critics. *Journal of Economic Issues, 51*(3), 588–612.

Tarde, G. (1903). *The laws of imitation*. transl. by E. C. Parsons. New York: Henry Holt.

Trompenaars, F., & Hampden-Turner, C. (2012). *Riding the waves of culture: Understanding diversity in global business* (3rd ed.). New York: McGraw-Hill.

Turner, J. H., Machalek, R., & Maryanski, A. (Eds.). (2015). *Handbook on evolution and society: Toward an evolutionary social science*. Abingdon: Routledge.

Tyler, T. (2011). *Memetics: Memes and the science of cultural evolution*. Mersenne Publishing.

Urmetzer, S., Schlaile, M. P., Bogner, K., Mueller, M., & Pyka, A. (2018). Exploring the dedicated knowledge base of a transformation towards a sustainable bioeconomy. *Sustainability, 10*(6), 1694. https://doi.org/10.3390/su10061694.

van den Bergh, J. C. J. M. (2018). *Human evolution beyond biology and culture: Evolutionary social, environmental and policy sciences*. Cambridge: Cambridge University Press.

Veblen, T. (1898). Why is economics not an evolutionary science? *The Quarterly Journal of Economics, 12*(4), 373. https://doi.org/10.2307/1882952.

Veblen, T. (1899). *The theory of the leisure class: An economic study of institutions*. New York: The Macmillan Company.

Weber, M. (1922/2013). *Economy and society: An outline of interpretive sociology* (G. Roth & C. Wittich, (Eds.)). Berkeley: University of California Press.

Weber, M. (1930). *The Protestant ethic and the spirit of capitalism*. Transl. by T. Parsons; with an introduction by A. Giddens. London: Routledge.

Wilson, D. S. (2002). *Darwin's cathedral: Evolution, religion, and the nature of society*. Chicago: University of Chicago Press.

Wilson, D. S. (2015). *Does altruism exist? Culture, genes, and the welfare of others*. New Haven: Yale University Press.

Wilson, D. S. (2019a). *Master class: A Conversation with Jonathan Birch about the equivalence of theories of social evolution*. The Evolution Institute. Retrieved from https://evolution-institute.org/master-class-a-conversation-with-jonathan-birch-about-the-equivalence-of-theories-of-social-evolution/.

Wilson, D. S. (2019b). *This view of life: Completing the Darwinian revolution*. New York: Pantheon Books.

Wilson, E. O. (1998). *Consilience: The unity of knowledge*. New York: Random House.

Witt, U. (2003). *The evolving economy: Essays on the evolutionary approach to economics*. Cheltenham: Edward Elgar.

Witt, U. (2014). The future of evolutionary economics: Why the modalities of explanation matter. *Journal of Institutional Economics*, *10*(4), 645–664. https://doi.org/10.1017/S1744137414000253.

Witt, U. (2016). *Rethinking economic evolution: Essays on economic change and its theory*. Cheltenham: Edward Elgar.

Wuketits, F. M. (1997). The philosophy of evolution and the myth of progress. *Ludus Vitalis*, *5*(9), 5–17.

Yates, T. (2019). *Memetics and economics*. Retrieved from https://longandvariable.wordpress.com/2019/05/08/memetics-and-economics/.

Zelizer, V. A. (2011). *Economic lives: How culture shapes the economy*. Princeton, NJ: Princeton University Press.

Chapter 2
"Meme Wars": A Brief Overview of Memetics and Some Essential Context

Michael P. Schlaile

Abstract This chapter provides an overview of memetics, starting with relevant terminology and examples of applications of memetics in other disciplines in Sect. 2.1. Section 2.2 highlights important contributions as well as cornerstones of the so-called "meme's eye view". Section 2.3 turns to some of the controversies in meme theory before Sects. 2.4–2.6 give an overview of the central notions of replicators, interactors, and memeplexes.

2.1 Episode I: The Phantom Memetics

In view of the broad and diverse literature, some remarks on the origins and cornerstones of memetics are indicated. However, due to the focus of this book, the following sections should be seen as a complement rather than a substitute for reading the literature reviews and discussions that have been published elsewhere.[1] Although the term "meme" was originally coined (in 1976) by Dawkins (2016, p. 249) as a "unit of cultural transmission, or a unit of *imitation*" (italics in original) by analogy with the gene, some of the central ideas of memetics have been around for much longer

---

[1] Concise overviews and introductions to memetics can be found, for example, in Aunger (2007), Baraghith (2015, Chap. 2), Burman (2012), Heylighen and Chielens (2009), Jesiek (2003), Kronfeldner (2011, Chaps. 5–6), Mick (2019, Chap. 1), Patzelt (2015b), Salwiczek (2001), Stewart-Williams (2018, Chap. 6 and pp. 293–303), and von Bülow (2013a, 2013b). For more general reviews of the literature on cultural evolution, see, for instance, Lewens (2019), Linquist (2010), Mesoudi (2007, 2015, 2017), or Reisman (2013).

The original version of this chapter was revised: The word "muchlonger" in page 15 was corrected as "much longer" and the word "several doctoraldissertations" in page 17 was corrected as "several doctoral dissertations". The correction to this chapter is available at https://doi.org/10.1007/978-3-030-59955-3_9

M. P. Schlaile (✉)
Department of Innovation Economics, University of Hohenheim, Stuttgart, Germany
e-mail: schlaile@uni-hohenheim.de

© The Author(s), under exclusive license to Springer Nature Switzerland AG 2021, corrected publication 2021
M. P. Schlaile (ed.), *Memetics and Evolutionary Economics*, Economic Complexity and Evolution, https://doi.org/10.1007/978-3-030-59955-3_2

(as already indicated by the quotes near the end of Sect. 1.1).[2] Indeed, Dawkins (2016, p. 247) himself acknowledges several sources of inspiration (aside from Charles Darwin, of course), including L.L. Cavalli-Sforza (1971), F.T. Cloak (1975), J.M. Cullen (1972), and Karl Popper (1974b, 1978). It should, therefore, not be too surprising that others have even traced the "genealogy" of the meme back to philosophers in Ancient Greece. For example, as Momme von Sydow elucidates:

> The many aspects of the concept of a meme have a much longer history than its new name, reaching back at least two and a half thousand years to the concept of *nous* (partly also to the concept of *logos*), presumably best translated as *spirit,* or in an individualistic sense as *mind* (von Sydow 2012, p. 57, italics in original).[3]

On a more sarcastic note, Alan Costall remarks that "Dawkins manages to leave us with the impression that he has come up with a pretty bright new idea" (Costall 1991, p. 323, with reference to Campbell 1974, Gerard et al. 1956, and others). Similarly, Tim Ingold (1986), who strongly opposes the notion of Cloak and Dawkins that cultural traits could be "parasitic," sees some roots of memes in the works of Franz Boas. At one point, he even calls memes "the bastard offspring of Boas's cultural elements" (Ingold 1986, p. 68, with reference to Boas 1940). And Franz Wegener (2015) reminds us in his introductory quote that another important forefather of memetics can be found in Leo Frobenius (1921):[4]

> The *Kulturkreis* school … [conceives of] culture as an organism independent from its human carriers [and] regards every cultural form as a living entity on its own … First of all: cultures are not brought forth by human will, culture rather lives 'on' humans (Frobenius 1921, pp. 3–4, own translation).

To be clear, this book is not about the intellectual history of memetics—although that would be an interesting study in its own right—but it is important to understand that, when it comes to memes, Dawkins' principal contribution was the recombination of ideas and concepts that had already been "in the meme pool," if you will,[5] and to have come up with a pretty catchy name. Why is this important? Because many of the arguments against memes appear to target Dawkins' admittedly ambiguous conception (see also Sukopp 2010) and tend to disregard much of the conceptual and empirical progress in the field. So, what is the status quo of meme theory, anyway?

Nowadays, one could get the impression that memes are found mainly on the Internet and in computer science: The term meme has become a synonym for funny

---

[2] Note that Dawkins has also contributed to the meme's conceptual ambiguity by redefining memes as "units of information residing in a brain" in *The Extended Phenotype* (Dawkins 1982b, p. 109) and by likening memes to viruses (Dawkins 1993). Yet, to be fair, Dawkins' introduction of the meme was never intended as a fully fledged theory of cultural evolution but to make the case that natural selection operates on *replicators* more generally, not "selfish genes" in particular (cf. Dawkins 1999, p. xvi). As Burman (2012, p. 77) puts it: "The original meme, in other words, was a rhetorical flourish intended to clarify a larger argument."

[3] See also Moritz (1990, pp. 6–8) on some of the philosophical roots of memetics.

[4] See also Jahoda (2002) or Marsden (2000), for comments on other precursors and antecedents of memetics.

[5] Note that this is the essence of the memetic take on creativity in any case (e.g., cf. Dennett 2017), as we will also briefly revisit in Sect. 7.2.

pictures, videos, texts, etc., which are shared rapidly by Internet users;[6] and if you search for academic publications with "memetic" in their title you will find countless publications on so-called *memetic algorithms*, an area of evolutionary computation (e.g., Baydin de Mántaras 2012; Moscato 1989; Neri and Cotta 2012; Neri et al. 2012). Given that there is also a distinct new research field focusing on memes in a digital context (e.g., Burgess et al. 2018; Johann and Bülow 2019; Segev et al. 2015; Shifman 2014; Wiggins and Bowers 2015), it almost looks as if memes have found a new niche in the digital world. However, this does not mean that the concept has become useless or meaningless in other contexts. Quite the contrary, enlightening examples of applied memetics include (but are not limited to) organizational memetics (e.g., Price 1995; Shepherd and McKelvey 2009; Taylor and Giroux 2005; Weeks and Galunic 2003; see also Chap. 4 below), spiral dynamics (Beck and Cowan 1996; Beck et al. 2018), translation theory (Chesterman 2000, 2005a, 2005b, 2009, 2016) or (the evolution of) language more generally (e.g., Simmel 2009; Worden 2004), biblical studies (Pyper 1998), marketing (e.g., Atadil et al. 2017; Hamlin et al. 2015; Marsden 1998, 2002; Murray et al. 2014; R. Williams 2000, 2002, 2004), contagion in financial markets (Lynch 2000; Hirshleifer and Teoh 2009), a memetic approach to the First Amendment (Stake 2001), discussions of memetic influences on markets and companies (e.g., Breitenstein 2002),[7] the evolution of religions (Dennett 2006),[8] the memetics of music (Jan 2007, 2016), the evolution of institutions (Patzelt 2007, 2012, 2015a, 2017), the study of history and myth (Amitay 2010), research on mental illness (Leigh 2010), the study of patented technology (Bedau 2013) and trademarks (Johnson 2013), the evolution and spread of information via social networks (e.g., Adamic et al. 2016; Barabási 2016, Chap. 10; Weng 2014; Weng et al. 2012), the analysis of memes in scientific literature (Kuhn et al. 2014), the analysis of "viral" Internet phenomena (e.g., Dobson and Sukumar 2017; Spitzberg 2014; see also Chap. 6 below), the identification and analysis of "units" in creative media (Velikovsky 2016, 2018) and especially television (Kneis 2010), or to explain the phenomenon of witchcraft persecutions in early modern Europe (Boudry and Hofhuis 2018; Hofhuis and Boudry 2019), and several others (see also Chap. 5 in Blute 2010, for more examples). In the light of (political or religious) ideologies and their potential influence on people's propensity for hostility, it is hardly surprising that memetics has also received attention from military, defense, security, and intelligence researchers (e.g., Finkelstein 2011; Finkelstein and Ayyub 2010; Fries and Singpurwalla 2008; Pech 2003; Prosser 2006; Prosser and Giscoppa 2013, Chap. 17). On top of this, memetic approaches have also inspired—or were embraced in—several doctoral dissertations, including quite recent ones (e.g., Cook 2008; Cullen 1995; Gill 2013; Karafiáth 2015; Nye

---

[6]To take up the *Oxford Dictionaries* again, the second definition of a meme given there reads: "An image, video, piece of text, etc., typically humorous in nature, that is copied and spread rapidly by Internet users, often with slight variations" (http://www.oxforddictionaries.com/definition/english/meme); see also Marwick (2013).

[7]Some authors (e.g., Mérő 2009) even consider money itself to be a distinct replicator or another kind of meme (Westoby 1994, Sect. "Derivative memes").

[8]For an adaptationist alternative to Dennett's memetic explanation, see, for example, Bulbulia (2008).

2011; Phuaphanthong 2014; Russ 2014; Shepherd 2002, Sparkes-Vian 2014; Weng 2014, to name but a few).

2.2 Episode II: Attack of the Meme Machines

From the early 1980s to the mid-1990s, both Douglas Hofstadter and Daniel Dennett made a case for memes in their widely read books (e.g., Dennett 1991, 1995; Hofstadter 1985; Hofstadter and Dennett 1981). Whereas Dennett's and Hofstadter's books were not specifically about memes, in 1996 two popular science books dedicated solely to this topic were published by Richard Brodie and Aaron Lynch (Brodie 1996; Lynch 1996). While the books and newspaper articles contributed to the circulation of the meme concept among the general public (see also Burman 2012, for details), a scientific audience took it up as well—both endorsing and criticizing it (e.g., Ball 1984; Cullen 1993, 1996; Delius 1989, 1991; Dennett 1990; Hull 1982).[9] Eventually, in 1997, the scientific study of memes gained a central outlet, the *Journal of Memetics—Evolutionary Models of Information Transmission*.

Another seminal contribution to the memetics of the late 1990s is undoubtedly Susan Blackmore's (1999b) book *The Meme Machine*. Blackmore takes up Dawkins' (1976) cue that memes are units of imitation and develops these arguments further. She also makes a strong case for the *meme's eye view* (Blackmore 2000, particularly following Dennett 1995), that is, the view that memes compete for (the attention of) human minds, while their successful propagation does not necessarily depend on their truth or usefulness for the human "vehicles" or "meme machines." Blackmore frequently stresses that speaking of "selfish" replicators in terms of active and intentional entities that want to be passed on is just a useful shorthand for a much more complicated evolutionary process (e.g., cf. Blackmore 1999b, p. 5). Yet, it is exactly this manner of speaking which is attractive to some (e.g., Boudry 2018a; Pagel 2012; Stanovich 2005) but also heavily criticized and frequently misunderstood by others. For example, as Patricia Williams writes:

> Blackmore turns memes into actors who have desires and interests. Repeatedly she says this is only a way of speaking, but like Dawkins before her, she tangles herself in her own language, trips, and falls into the pit. ... Blackmore ... [also] mimics him in making people passive. She reverses actor and object in this manner ... In her book, memes are rulers; people are passive hosts (as in parasite and host) or vehicles (as in Dawkins) (P. Williams 2002, p. 305).

On that note, Tim Lewens (2015) also expresses his doubts as to whether we need the meme's eye view to explain "various forms of irrationality, weakness of will, self-deception, false consciousness, subconsciously motivated action, and so forth"

---

[9]In fact, some authors speculate that the reluctance of many researchers to take up the term meme is due not only to its terminological ambiguity but also to its popularity in nonacademic circles (e.g., Knobloch 2015; Lord 2012).

(p. 31). Nevertheless, in a quite recent defense of the meme's eye view, Maarten Boudry directly responds to Lewens' objection in the following way:

> Lewens is right that psychologists and behavioral economists already have plenty of resources to explain various sorts of human irrationality ... We don't need the memeticists to tell us that. But memetics – or the meme's eye view – is helpful to explain the evolution of complex *systems* of misbeliefs (Boudry 2018a, p. 119, italics in original).

However, Boudry criticizes—and quite rightfully so, one might add—a worldview he calls "panmemetics," that is, the overzealous claims of some meme enthusiasts that (the evolution of) *all* cultural phenomena can be explained by means of selfish memes (Boudry 2018a; Boudry and Hofhuis 2018).[10]

2.3 Episode III: Revenge of the Myth

It did not take very long for the controversies among meme enthusiasts themselves (e.g., Blute 2005; Gatherer 1998b; Hull 1999; Lynch 1999; Marsden 1999; Rose 1998) and also the critics with an already more suspicious attitude toward memes (e.g., Atran 2001; Aunger 2000; Edmonds 2002, 2005; Jahoda 2002; McGrath 2004, Chap. 4; Sperber 1996) to become more numerous, more serious, and eventually "successful" in conducing to the termination of the *Journal of Memetics* in 2005. However, as the research contributions mentioned at the end of Sect. 2.1 show, the demise of the journal did not mark the end of scientific studies taking up (at least some version of) memetics. But while the concept found its way into studies from many different disciplines, the interest of evolutionary economists and cultural evolutionists in memes began to wane, as also Kevin Laland (2017) underlines by noting that "the modern science of cultural evolution derives very little from memetics" (p. 323). However, keeping the research question of this book in mind, it should be noted that various evolutionary economists—at least at some point in time—seem to have adopted memes into their vocabulary.[11]

In the meantime, cultural evolutionists seem to have (re-)turned to (connotation-wise) more harmless—though not necessarily less ambiguous—terms such as "cultural traits," "cultural variants," etc. (e.g., Boyd and Richerson 2005; Richerson and Boyd 2005),[12] whereas some evolutionary economists have developed their approaches by drawing on concepts and terms like "habits," "routines," "rules," "knowledge units," etc. (we will return to this issue below in Chap. 3).

---

[10] On a related note, Lawson (2003) explains his hesitation to embrace memetics mainly with its reductionist "view that the replicator is the prime mover *in all that happens*" (p. 135, emphasis added).

[11] Examples include but are not limited to Arthur (cf. 2009, p. 102), Binmore (1998), Herrmann-Pillath (2000, Chap. III; 2010, Chap. 3), and some chapters in Ziman (2000).

[12] And instead of "parasitic memes" they simply use more magniloquent expressions like "maladaptive cultural variants" (e.g., Richerson and Boyd 2005, Chap. 5) or "maladaptive culturally transmitted traits" (e.g., Boyd 2018, p. 182).

Although it is quite understandable that serious researchers want to disassociate themselves from the more esoteric and mystical associations inherent in (pan-)memetic terminology, it neither makes the underlying theoretical issues go away nor helps to achieve *consilience*. In other words, having a common language could arguably be useful for avoiding reinventing the wheel in different scientific disciplines that deal with similar questions and research objects but use different terminology. Speaking of terminology, in the following two subsections, we will, therefore, briefly revisit two central concepts in memetics and other Darwinian approaches, namely, *replicators* and *interactors* (see also Aldrich et al. 2008, esp. pp. 586–588).

2.4 Episode IV: A New Replicator

Arguably, one of the most central concepts in Dawkins' (1976) *The Selfish Gene* is that of the *replicator*—of which both genes and memes are said to be instances:

> The fundamental units of natural selection, the basic things that survive or fail to survive, that form lineages of identical copies with occasional random mutations, are called replicators (Dawkins 2016, p. 328).

The term is specified by Dawkins in a published version of a lecture given in 1977:

> We may define a replicator as any entity in the universe which interacts with its world, including other replicators, in such a way that copies of itself are made. A corollary of the definition is that at least some of these copies, in their turn, serve as replicators, so that a replicator is, at least potentially, an ancestor of an indefinitely long line of identical descendant replicators (Dawkins 1978, p. 67).

In the foreword to Blackmore (1999b), Dawkins (1999) simplifies but also amends this definition to "any unit of which copies are made, with occasional errors, and *with some influence or power over their own probability of replication*" (p. xvi, emphasis added). Note that the italicized part—the influence or power over their own replicative success—is what constitutes an "active" replicator in Dawkins' view.[13] Rather predictably, the notion of active replicators has evoked some objections against memes as well. However, concerning this matter, Douglas Roy presents the following counterargument:

> [T]hese critics seem to be arguing that memes might be 'passive replicators' in that they get copied, but cannot be 'active replicators,' which are selected. The fact, though, that such minimalist replicators as viruses evolve—without having their own cells or metabolism—should give pause to such arguments. Many varieties of replicators have been described that are demonstrably subject to natural selection without creating any of the biological machinery critics of memes assert is essential for memes to be a viable concept (Roy 2017, p. 286).

---

[13] In Dawkins (1982a), the differences between so-called germ-line and dead-end replicators as well as between active and passive replicators are explained, but we will not get caught up in these details, here. Moreover, it should be noted that according to Dawkins (2016), "successful" replicators exhibit three important characteristics: copying-fidelity, fecundity, and longevity (see also von Buelow 2013b, for a short review).

Notably, the replicator concept was taken up and promoted by David Hull quite early on. Hull, who also had some influence on the generalized Darwinian school of thought in evolutionary economics (e.g., as one of the coauthors of Aldrich et al. 2008), explains:

> Certain entities (replicators) pass on their structure largely intact from generation to generation. These entities either interact with their environments in such a way as to bias their distribution in later generations or else produce more inclusive entities that do. As a result, even more inclusive entities evolve (Hull 1980, p. 315).

According to the proponents of a generalized Darwinism in evolutionary economics, the *process* of replication necessarily involves causality, similarity, and information transfer (cf. Aldrich et al. 2008, p. 586; Almudi and Fatas-Villafranca 2018, p. 84; Hodgson and Knudsen 2010a, p. 77). Unsurprisingly, there has also been some debate about these three criteria or the notion and importance of replicators for Darwinian evolutionary processes more generally.[14]

Without losing ourselves in these ongoing debates, we shall now briefly turn to the aspect of information (transfer), which is of central importance for the chapters that follow. As already John Ball reminds us:

> The essence of a replicable gene or meme is information; indeed, I would argue that genes and memes *are* information and that DNA, RNA, paper, and so forth are *media* on which replicators are written or by which replicators are reproduced or communicated (Ball 1984, p. 154, italics in original).

In the exact same manner, George Williams (1986) writes in his comments on Elliott Sober's (1984) book *The Nature of Selection*:

> A gene is ... a weightless package of information that plays an instructional role in development. An analogy in cultural evolution would be a popular song ... It is a transmittable package of information that controls singing behavior. It can be coded as molecular subtleties in brains, as magnetic patterns on tape, ink on paper, etc. It competes with other songs just as genes compete with genes, because only a limited amount of singing can be performed, and because different kinds of singing are perceived to have different musical fitness in the prevailing cultural environment (Williams 1986, p. 121).

This sentiment is also echoed by Dennett:

> Memes—cultural recipes—similarly depend on one physical medium or another for their continued existence (they aren't magic), but they can leap around from medium to medium, being translated from language to language, from language to diagram, from diagram to rehearsed practice, and so forth (Dennett 2002, p. E-88).

Of course, the notion of *information* itself is not unproblematic (e.g., Hodgson and Knudsen 2008, 2010b; Lewens 2015, Chap. 3),[15] but Boudry (2018b) certainly has a point when he writes:

---

[14] Recent overviews and introductions to the debate(s) about replicators in biology can, for example, be found in Godfrey-Smith (2000), Lloyd (2017), Wilkins and Bourrat (2019), and more extensively in Jablonka and Lamb (2014).

[15] For example, see also Magne's (2015, 2016) objection to Dennett's informational approach to memes.

If you identify cultural representations with some sort of physical substrate, you will look in vain for some identifiable replication mechanism that makes copies of those material entities, in the way that DNA strands unwind and assemble a replica. But on an informational approach, this is not necessary. What matters is that some piece of information is somehow transmitted from one physical carrier to the next, no matter how roundabout and circuitous the mechanism, and no matter how different the physical carriers (Boudry 2018b, p. 27).

2.5 Episode V: The Interactor Strikes Back

While Dawkins (1976) introduces the notion of the replicator, he also proposes the counterpart "vehicle" to denote the entity that contains the replicator. Vehicle is defined by Dawkins (1982b, p. 114) as "any unit, discrete enough to seem worth naming, which houses a collection of replicators and which works as a unit for the preservation and propagation of these replicators." In the case of the gene, vehicles are usually individual organisms, but Dawkins also notes that there can be hierarchies of nested vehicles (Dawkins 1982a). Blackmore (1999b) follows Dawkins in adopting a rather passive terminology of vehicles and machines, but she also mentions Hull's alternative notion of the "interactor," which has a more active connotation (cf. Hull 1988a).[16] Hull writes:

> Memes can exist in brains, books, computers, and a wide variety of physical vehicles of knowledge.... Mental telepathy notwithstanding, memes cannot be transferred directly from one brain to another. Some sort of physical intermediary is necessary (Hull 1982, p. 276).

Hull calls this physical intermediary the *interactor*, which is defined as "an entity that interacts as a cohesive whole with its environment in such a way that this interaction *causes* replication to be differential" (Hull 1988b, p. 408, italics in original). However, this notion has also caused some confusion, which is easy to see if we look back at the replicator definitions in the previous section and notice that both Dawkins and Hull at some point argue that a replicator *interacts* with "its world" or its "environment."[17] The observation that the distinction between replicators and interactors may become blurred is also supported by Hans-Cees Speel's argument that memes can be interactors as well—especially in terms of "internal" or selective interaction inside a mind (Speel 1998).

This ambiguity becomes somewhat problematic also for the generalized Darwinian approach to evolutionary economics, especially since Hodgson and Knudsen (2010a, p. 24) follow Robert Brandon's (1996, p. 125) advice that the "distinction between replicators and interactors ... is best seen as a generalization of the traditional genotype-phenotype distinction." We can see that this "generalization" might not always hold in the socioeconomic realm, as also Hull (1982, p. 307) suggests by

---

[16] It should be noted, however, that Hull himself does not regard Dawkins' vehicles and his interactors as perfectly equivalent concepts (cf. Hull 1988a, p. 31).

[17] For example, Hull (1988a, p. 31) also writes: "On my account, genes are both replicators and interactors. If genes are anything, they are entities that interact with their environments in such a way as to bias their own replication."

writing that "the genotype/phenotype interface, which is so important in biological evolution, is not all that apparent in memetic evolution."

Against the background of these and other issues, several proponents of a memetic approach have taken a different road that is inspired, for example, by Cloak's (1975) distinction between i-culture and m-culture, McNamara's (2011) related classification of memes into i-memes and e-memes, and especially Popper's *three worlds* (e.g., 1972, 1974a). Authors such as Chesterman (2016), Gatherer (1998a), Patzelt (2015b), Schlaile and Ehrenberger (2016), and Wegener (2015) argue for an approach to memes that attempts to bypass some of the issues connected to the replicator/interactor dichotomy mentioned above. It would go well beyond the scope of this book to delve into the details, differences, and difficulties of these approaches—especially because we could easily get sucked into discussions about philosophy of mind and the mind–body problem that can quickly get out of hand. We will briefly return to the three-dimensional approach proposed by myself and Marcus Ehrenberger (2016) below in Chap. 3. Nevertheless, the unresolved philosophical issues will have to remain an important topic for future research on memes.

2.6 Episode VI: Return of the Memeplex

The final concept addressed in this overview chapter is that of the *memeplex*. Whereas Dawkins (1976) originally uses the rather intricate term "co-adapted meme-complexes" (e.g., Dawkins 2016, p. 258), Blackmore (1999b) follows Speel (1999) and abbreviates the term to memeplexes. As Dawkins (1999) explains:

> Memes, like genes, are selected against the background of other memes in the meme pool. The result is that gangs of mutually compatible memes—coadapted meme complexes or memeplexes—are found cohabiting in individual brains (Dawkins 1999, p. xiv).

According to Blackmore (1999a, p. 44), examples of memeplexes "include languages, religions, scientific theories, political ideologies and belief systems such as acupuncture or astrology." To take up another passage from Dawkins (1982b) (but see also Distin 2005, for similar arguments):

> It is true that the relative survival success of a meme will depend critically on the social and biological climate in which it finds itself, and this climate will certainly be influenced by the genetic make-up of the population. But it will also depend on the memes that are already numerous in the meme pool. … [A]ny new meme's replication success will be influenced by its compatibility with this existing background (Dawkins 1982b, p. 111).

Francis Heylighen and Klaas Chielens (2009) also use the term memeplex to denote a "collection of mutually supporting memes, which tend to replicate together" (p. 3205), and they propose several criteria that can, in principle, explain the replicative success of memes and their memeplexes. We will return to the issues of memeplexes and compatibility-based meme transmission from a theoretical perspective in Sect. 3.3 and as a central theme of the simulation model in Chap. 5.

References

Adamic, L. A., Lento, T. M., Adar, E., & Ng, P. C. (2016). Information evolution in social networks. In P. N. Bennett, V. Josifovski, J. Neville, & F. Radlinski (Eds.), *Proceedings of the 9th ACM international conference on web search and data mining - WSDM'16* (pp. 473–482). New York: ACM Press.

Aldrich, H. E., Hodgson, G. M., Hull, D. L., Knudsen, T., Mokyr, J., & Vanberg, V. J. (2008). In defence of generalized Darwinism. *Journal of Evolutionary Economics, 18*, 577–596. https://doi.org/10.1007/s00191-008-0110-z.

Almudi, I., & Fatas-Villafranca, F. (2018). Promotion and coevolutionary dynamics in contemporary capitalism. *Journal of Economic Issues, 52*(1), 80–102. https://doi.org/10.1080/00213624.2018.1430943.

Amitay, O. (2010). *From Alexander to Jesus*. Berkeley, CA: University of California Press.

Arthur, W. B. (2009). *The nature of technology: What it is and how it evolves*. New York: Free Press.

Atadil, H. A., Sirakaya-Turk, E., Baloglu, S., & Kirillova, K. (2017). Destination neurogenetics: Creation of destination meme maps of tourists. *Journal of Business Research, 74*, 154–161. https://doi.org/10.1016/j.jbusres.2016.10.028.

Atran, S. (2001). The trouble with memes: Inference versus imitation in cultural creation. *Human Nature, 12*(4), 351–381. https://doi.org/10.1007/s12110-001-1003-0.

Aunger, R. (Ed.). (2000). *Darwinizing culture: The status of memetics as a science*. Oxford: Oxford University Press.

Aunger, R. (2007). Memes. In R. I. M. Dunbar & L. Barrett (Eds.), *The Oxford handbook of evolutionary psychology* (pp. 599–604). Oxford: Oxford University Press.

Ball, J. A. (1984). Memes as replicators. *Ethology and Sociobiology, 5*, 145–161. https://doi.org/10.1016/0162-3095(84)90020-7.

Barabási, A.-L. (2016). *Network science*. Cambridge: Cambridge University Press.

Baraghith, K. (2015). *Kulturelle Evolution und die Rolle von Memen: Ein Mehrebenenmodell*. Frankfurt a. M.: Peter Lang.

Baydin, A. G., & de Mántaras, R. L. (2012). Evolution of ideas: A novel memetic algorithm based on semantic networks. In *IEEE Congress on Evolutionary Computation (CEC), 2012* (pp. 1–8). https://doi.org/10.1109/CEC.2012.6252886

Beck, D. E., & Cowan, C. C. (1996). *Spiral dynamics: Mastering values, leadership, and change - Exploring the new science of memetics*. Malden, MA: Blackwell.

Beck, D. E., Larsen, T. H., Solonin, S., Viljoen, R. C., & Johns, T. Q. (2018). *Spiral dynamics in action: Humanity's master code*. Chichester: Wiley.

Bedau, M. A. (2013). Minimal memetics and the evolution of patented technology. *Foundations of Science, 18*(4), 791–807. https://doi.org/10.1007/s10699-012-9306-7.

Binmore, K. (1998). *Game theory and the social contract II: Just playing*. Cambridge, MA: The MIT Press.

Blackmore, S. (1999a). Meme, myself, I. *New Scientist, 161*(2177), 40–44.

Blackmore, S. (1999b). *The meme machine*. Oxford: Oxford University Press.

Blackmore, S. (2000). The meme's eye view. In R. Aunger (Ed.), *Darwinizing culture: The status of memetics as a science* (pp. 25–42). Oxford: Oxford University Press.

Blute, M. (2005). Memetics and evolutionary social science. *Journal of Memetics - Evolutionary Models of Information Transmission, 9*. Retrieved from http://cfpm.org/jom-emit/2005/vol9/blute_m.html.

Blute, M. (2010). *Darwinian sociocultural evolution: Solutions to dilemmas in cultural and social theory*. Cambridge: Cambridge University Press.

Boas, F. (1940). *Race, language and culture*. New York: Macmillan.

Boudry, M. (2018a). Invasion of the mind snatchers. On memes and cultural parasites. *Teorema, 37*(2), 111–124.

Boudry, M. (2018b). Replicate after reading: On the extraction and evocation of cultural information. *Biology & Philosophy, 33*. https://doi.org/10.1007/s10539-018-9637-z.

Boudry, M., & Hofhuis, S. (2018). Parasites of the mind. Why cultural theorists need the meme's eye view. *Cognitive Systems Research, 52*, 155–167. https://doi.org/10.1016/j.cogsys.2018.06.010.

Boyd, R. (2018). *A different kind of animal: How culture transformed our species*. Princeton: Princeton University Press.

Boyd, R., & Richerson, P. J. (2005). *The origin and evolution of cultures*. Oxford: Oxford University Press.

Brandon, R. N. (1996). *Concepts and methods in evolutionary biology*. Cambridge: Cambridge University Press.

Breitenstein, R. (2002). *Memetik und Ökonomie: Wie die Meme Märkte und Organisationen bestimmen*. Münster, Germany: Lit.

Brodie, R. (1996). *Virus of the mind. The new science of the meme*. Seattle: Integral Press.

Bulbulia, J. (2008). Meme infection or religious niche construction? An adaptionist alternative to the cultural maladaptationist hypothesis. *Method and Theory in the Study of Religion, 20*, 67–107. https://doi.org/10.1163/157006808X260241.

Burgess, A., Miller, V., & Moore, S. (2018). Prestige, performance and social pressure in viral challenge memes: Neknomination, the ice-bucket challenge and smearforsmear as imitative encounters. *Sociology, 52*(5), 1035–1051. https://doi.org/10.1177/0038038516680312.

Burman, J. T. (2012). The misunderstanding of memes: Biography of an unscientific object, 1976–1999. *Perspectives on Science, 20*(1), 75–104. https://doi.org/10.1162/POSC_a_00057.

Campbell, D. T. (1974). Unjustified variation and selective retention in scientific discovery. In F. J. Ayala & T. Dobzhansky (Eds.), *Studies in the philosophy of biology* (pp. 139–161). London: The Macmillan Press.

Cavalli-Sforza, L. L. (1971). Similarities and dissimilarities of sociocultural and biological evolution. In F. R. Hodson, D. Kendall, & P. Tautu (Eds.), *Mathematics in the archaeological and historical sciences* (pp. 535–541). Edinburgh: Edinburgh University Press.

Chesterman, A. (2000). Memetics and translation studies. *Synaps, 5*, 1–17.

Chesterman, A. (2005a). Consilience in translation studies. *Revista Canaria de Estudios Ingleses, 51*, 19–32.

Chesterman, A. (2005b). The memetics of knowledge. In H. van Dam, J. Engberg, & H. Gerzymisch-Arbogast (Eds.), *Knowledge systems and translation* (pp. 17–30). Text, translation, computational processing. Berlin: de Gruyter.

Chesterman, A. (2009). The view from memetics. *Paradigmi. Rivista di critica filosofica, 2*, 75–88.

Chesterman, A. (2016). *Memes of translation: The spread of ideas in translation theory* (Revised ed.). Amsterdam: John Benjamins.

Cloak, F. T. (1975). Is a cultural ethology possible? *Human Ecology, 3*(3), 161–182. https://doi.org/10.1007/BF01531639.

Cook, J. E. (2008). *The role of the individual in organisational cultures: A Gravesian integrated approach*. Doctoral dissertation, Sheffield Hallam University.

Costall, A. (1991). The 'meme' meme. *Cultural Dynamics, 4*(3), 321–335. https://doi.org/10.1177/092137409100400305.

Cullen, B. (1993). The Darwinian resurgence and the cultural virus critique. *Cambridge Archaeological Journal, 3*(2), 179–202. https://doi.org/10.1017/S0959774300000834.

Cullen, B. (1995). *The cultural virus*. Doctoral dissertation, School of Archaeology, Classics, and Ancient History. Sydney: University of Sydney.

Cullen, B. (1996). Cultural virus theory and the eusocial pottery assemblage. In H. D. G. Maschner (Ed.), *Darwinian archaeologies* (pp. 43–59). New York: Plenum Press.

Cullen, J. M. (1972). Some principles of animal communication. In R. A. Hinde (Ed.), *Non-verbal communication* (pp. 101–125). Cambridge: Cambridge University Press.

Dawkins, R. (1976). *The selfish gene*. Oxford: Oxford University Press.

Dawkins, R. (1978). Replicator selection and the extended phenotype. *Zeitschrift für Tierpsychologie, 47*(1), 61–76. https://doi.org/10.1111/j.1439-0310.1978.tb01823.x.
Dawkins, R. (1982a). Replicators and vehicles. In King's College Sociobiology Group (Ed.), *Current problems in sociobiology* (pp. 45–64). Cambridge: Cambridge University Press.
Dawkins, R. (1982b). *The extended phenotype: The long reach of the gene*. Oxford: Oxford University Press.
Dawkins, R. (1993). Viruses of the mind. In B. Dahlbohm (Ed.), *Dennett and his critics: Demystifying minds* (pp. 13–27). Oxford: Blackwell.
Dawkins, R. (1999). Foreword. In S. Blackmore (1999b), *The meme machine* (pp. vii–xvii). Oxford: Oxford University Press.
Dawkins, R. (2016). *The selfish gene* (40th anniversary ed.). Oxford: Oxford University Press.
Delius, J. D. (1989). Of mind memes and brain bugs; a natural history of culture. In W. A. Koch (Ed.), *The nature of culture* (pp. 26–79). Bochum: Studienverlag Brockmeyer. Retrieved from http://nbn-resolving.de/urn:nbn:de:bsz:352-206607.
Delius, J. D. (1991). The nature of culture. In M. S. Dawkins, T. R. Halliday, & R. Dawkins (Eds.), *The Tinbergen legacy* (pp. 75–99). London: Chapman & Hall.
Dennett, D. C. (1990). Memes and the exploitation of imagination. *The Journal of Aesthetics and Art Criticism, 48*(2), 127–135. https://doi.org/10.2307/430902.
Dennett, D. C. (1991). *Consciousness explained*. Boston (MA): Little Brown.
Dennett, D. C. (1995). *Darwin's dangerous idea: Evolution and the meanings of life*. London: Simon & Schuster.
Dennett, D. C. (2002). The new replicators. In M. Pagel (Ed.), *Encyclopedia of evolution* (pp. E83–E92). Oxford: Oxford University Press.
Dennett, D. C. (2006). *Breaking the spell: Religion as a natural phenomenon*. New York: Penguin.
Dennett, D. C. (2017). Memes 101: How cultural evolution works. *BigThink*. Retrieved from https://bigthink.com/videos/daniel-dennett-memes-101.
Distin, K. (2005). *The selfish meme: A critical reassessment*. Cambridge: Cambridge University Press.
Dobson, S., & Sukumar, A. (2017). Memes and civic action: Building and sustaining civic empowerment through the Internet. In C. Yamu, A. Poplin, O. Devisch, & G. De Roo (Eds.), *The virtual and the real in planning and urban design: Perspectives, practices and applications* (pp. 267–278). London: Routledge.
Edmonds, B. (2002). Three challenges for the survival of memetics. *Journal of Memetics - Evolutionary Models of Information Transmission, 6*. Retrieved from http://cfpm.org/jom-emit/2002/vol6/edmonds_b_letter.html.
Edmonds, B. (2005). The revealed poverty of the gene-meme analogy - why memetics per se has failed to produce substantive results. *Journal of Memetics - Evolutionary Models of Information Transmission, 9*. Retrieved from http://cfpm.org/jom-emit/2005/vol9/edmonds_b.html.
Finkelstein, R. (2011). *Tutorial: Military memetics*. Retrieved from http://www.robotictechnologyinc.com/images/upload/file/Presentation%20Military%20Memetics%20Tutorial%2013%20Dec%2011.pdf.
Finkelstein, R., & Ayyub, B. M. (2010). Memetics for threat reduction in risk management. In J. G. Voeller (Ed.), *Wiley handbook of science and technology for homeland security* (pp. 301–309). Hoboken: Wiley.
Fries, A., & Singpurwalla, N. D. (2008). Memetics - Overview and baseline models. IDA Document D-3599. Institute for Defense Analyses: Alexandria.
Frobenius, L. (1921). *Paideuma: Umrisse einer Kultur- und Seelenlehre*. Munich: C.H. Beck'sche Verlagsbuchhandlung.
Gatherer, D. (1998a). Meme pools, world 3, and Averroës's vision of immortality. *Zygon, 33*(2), 203–219. https://doi.org/10.1111/0591-2385.00141.
Gatherer, D. (1998b). Why the 'Thought Contagion' metaphor is retarding the progress of memetics. *Journal of Memetics - Evolutionary Models of Information Transmission, 2*. Retrieved from http://cfpm.org/jom-emit/1998/vol2/gatherer_d.html.

Gerard, R. W., Kluckhohn, C., & Rapoport, A. (1956). Biological and cultural evolution: Some analogies and explorations. *Behavioral Science, 1*(1), 6–34. https://doi.org/10.1002/bs.3830010103.

Gill, J. (2013). *Evaluating memetics: A case of competing perspectives at an SME*. Doctoral dissertation, Sheffield Hallam University.

Godfrey-Smith, P. (2000). The replicator in retrospect. *Biology & Philosophy, 15*(3), 403–423. https://doi.org/10.1023/A:1006704301415.

Hamlin, R., Bishop, D., & Mather, D. W. (2015). 'Marketing earthquakes': A process of brand and market evolution by punctuated equilibrium. *Marketing Theory, 15*(3), 1–22. https://doi.org/10.1177/1470593115572668.

Herrmann-Pillath, C. (2000). *Evolution von Wirtschaft und Kultur: Bausteine einer transdisziplinären Methode*. Marburg: Metropolis.

Herrmann-Pillath, C. (2010). *The economics of identity and creativity: A cultural science approach*. New Brunswick: Transaction.

Heylighen, F., & Chielens, K. (2009). Evolution of culture, memetics. In R. A. Meyers (Ed.), *Encyclopedia of complexity and systems science* (pp. 3205–3220). New York: Springer.

Hirshleifer, D., & Teoh, S. H. (2009). Thought and behavior contagion in capital markets. In T. Hens & K. R. Schenk-Hoppé (Eds.), *Handbook of financial markets: Dynamics and evolution* (pp. 1–56). Amsterdam: Elsevier.

Hodgson, G. M., & Knudsen, T. (2008). Information, complexity and generative replication. *Biology & Philosophy, 23*(1), 47–65. https://doi.org/10.1007/s10539-007-9073-y.

Hodgson, G. M., & Knudsen, T. (2010a). *Darwin's conjecture: The search for general principles of social and economic evolution*. Chicago: University of Chicago Press.

Hodgson, G. M., & Knudsen, T. (2010b). Generative replication and the evolution of complexity. *Journal of Economic Behavior & Organization, 75*(1), 12–24. https://doi.org/10.1016/j.jebo.2010.03.008.

Hofhuis, S., & Boudry, M. (2019). 'Viral' hunts? A cultural Darwinian analysis of witch persecutions. *Cultural Science Journal, 11*(1), 13–29. https://doi.org/10.5334/csci.116.

Hofstadter, D. R. (1985). *Metamagical themas: Questing for the essence of mind and pattern*. New York: Basic Books.

Hofstadter, D. R., & Dennett, D. C. (1981). *The mind's I: Fantasies and reflections on self and soul*. New York: Basic Books.

Hull, D. L. (1980). Individuality and selection. *Annual Review of Ecology and Systematics, 11*(1), 311–332. https://doi.org/10.1146/annurev.es.11.110180.001523.

Hull, D. L. (1982). The naked meme. In H. C. Plotkin (Ed.), *Learning, development, and culture: Essays in evolutionary epistemology* (pp. 273–327). Chichester: Wiley.

Hull, D. L. (1988a). Interactors versus vehicles. In H. C. Plotkin (Ed.), *The role of behavior in evolution* (pp. 19–50). Cambridge, MA: MIT Press.

Hull, D. L. (1988b). *Science as a process: An evolutionary account of the social and conceptual development of science*. Chicago: The University of Chicago Press.

Hull, D. L. (1999). Strategies in meme theory - a commentary on Rose's paper: Controversies in meme theory. *Journal of Memetics - Evolutionary Models of Information Transmission, 3*. Retrieved from http://cfpm.org/jom-emit/1999/vol3/hull_dl.html.

Ingold, T. (1986). *Evolution and social life*. Cambridge: Cambridge University Press.

Jablonka, E., & Lamb, M. J. (2014). *Evolution in four dimensions: Genetic, epigenetic, behavioral, and symbolic variation in the history of life* (2nd ed.). Cambridge, MA: MIT Press.

Jahoda, G. (2002). The ghosts in the meme machine. *History of the Human Sciences, 15*(2), 55–68. https://doi.org/10.1177/0952695102015002126.

Jan, S. (2007). *The memetics of music: A neo-Darwinian view of musical structure and culture*. Aldershot: Ashgate.

Jan, S. (2016). A memetic analysis of a phrase by Beethoven: Calvinian perspectives on similarity and lexicon-abstraction. *Psychology of Music, 44*(3), 443–465. https://doi.org/10.1177/0305735615576065.

Jesiek, B. K. (2003). *Betwixt the popular and academic: The histories and origins of memetics*. Master's thesis, Virginia Polytechnic Institute and State University, Blacksburg. Retrieved from http://hdl.handle.net/10919/42774.

Johann, M., & Bülow, L. (2019). One does not simply create a meme: Conditions for the diffusion of Internet memes. *International Journal of Communication, 13*, 1720–1742.

Johnson, S. J. (2013). Memetic theory, trademarks & the viral meme mark. *The John Marshall Review of Intellectual Property Law, 13*(1), 96–129.

Karafiáth, B. L. (2015). *Memetic marketing. Memetic researches and analyses in Hungarian corporate context*. Doctoral dissertation, Corvinus University of Budapest. https://doi.org/10.14267/phd.2015025.

Kneis, P. (2010). *The emancipation of the soul: Memes of destiny in American mythological television*. Frankfurt a. M.: Peter Lang.

Knobloch, C. (2015). Der Kulturbegriff der neoevolutionistischen Kulturkritik. In F. Deus, A.-L. Dießelmann, L. Fischer, & C. Knobloch (Eds.), *Die Kultur des Neoevolutionismus* (pp. 225–272). Bielefeld: Transcript.

Kronfeldner, M. (2011). *Darwinian creativity and memetics*. Durham: Acumen.

Kuhn, T., Perc, M., & Helbing, D. (2014). Inheritance patterns in citation networks reveal scientific memes. *Physical Review X, 4*, 041036. https://doi.org/10.1103/PhysRevX.4.041036.

Laland, K. N. (2017). *Darwin's unfinished symphony: How culture made the human mind*. Princeton: Princeton University Press.

Lawson, T. (2003). *Reorienting economics*. London: Routledge.

Leigh, H. (2010). *Genes, memes, culture, and mental illness: Toward an integrative model*. New York: Springer.

Lewens, T. (2015). *Cultural evolution: Conceptual challenges*. Oxford: Oxford University Press.

Lewens, T. (2019). Cultural evolution. In E. N. Zalta (Ed.), *The Stanford encyclopedia of philosophy (Summer 2019)*. Retrieved from https://plato.stanford.edu/archives/sum2019/entries/evolution-cultural/.

Linquist, S. (2010). Introduction. In S. Linquist (Ed.), *The evolution of culture* (pp. xi–xli). Farnham: Ashgate.

Lloyd, E. (2017). Units and levels of selection. In E. N. Zalta (Ed.), *The Stanford encyclopedia of philosophy, Summer 2017 edition*. Retrieved from https://plato.stanford.edu/archives/sum2019/entries/selection-units/.

Lord, A. S. (2012). Reviving organisational memetics through cultural Linnæanism. *International Journal of Organizational Analysis, 20*(3), 349–370. https://doi.org/10.1108/19348831211254143

Lynch, A. (1996). *Thought contagion: How belief spreads through society. The new science of memes*. New York: Basic Books.

Lynch, A. (1999). Thought contagion: How belief spreads through society. A response to Paul Marsden. *Journal of Artificial Societies and Social Simulation, 2* (3–4). Retrieved from http://jasss.soc.surrey.ac.uk/2/3/lynch.html.

Lynch, A. (2000). Thought contagions in the stock market. *Journal of Psychology and Financial Markets, 1*(1), 10–23. https://doi.org/10.1207/S15327760JPFM0101_03.

Magne, S. (2015). *Objections to Daniel Dennett's informational meme*. Retrieved from http://memelogic.blogspot.com/2015/11/informational-memes.html.

Magne, S. (2016). *G. C. Williams and clarifications about information*. Retrieved from http://memelogic.blogspot.com/2016/05/g-c-williams-and-confusion-about.html.

Marsden, P. (1998). Memetics: A new paradigm for understanding customer behaviour and influence. *Marketing Intelligence & Planning, 16*(6), 363–368. https://doi.org/10.1108/EUM0000000004541.

Marsden, P. (1999). Book review of Aaron Lynch: Thought contagion: How belief spreads through society. *Journal of Artificial Societies and Social Simulation, 2*(2). Retrieved from http://jasss.soc.surrey.ac.uk/2/2/review4.html.

Marsden, P. (2000). Forefathers of memetics: Gabriel Tarde and the laws of imitation. *Journal of Memetics - Evolutionary Models of Information Transmission, 4*. Retrieved from http://cfpm.org/jom-emit/2000/vol4/marsden_p.html.

Marsden, P. (2002). Brand positioning: Meme's the word. *Marketing Intelligence & Planning, 20*(5), 307–312. https://doi.org/10.1108/02634500210441558.

Marwick, A. (2013). Memes. *Contexts, 12*(4), 12–13. https://doi.org/10.1177/1536504213511210.

McGrath, A. E. (2004). *Dawkins' God: Genes, memes, and the meaning of life*. Malden, MA: Blackwell.

McNamara, A. (2011). Can we measure memes? *Frontiers in Evolutionary Neuroscience, 3*. https://doi.org/10.3389/fnevo.2011.00001.

Mérő, L. (2009). *Die Biologie des Geldes: Darwin und der Ursprung der Ökonomie*. Reinbek bei Hamburg: Rowohlt.

Mesoudi, A. (2007). Biological and cultural evolution: Similar but different. *Biological Theory, 2*(2), 119–123. https://doi.org/10.1162/biot.2007.2.2.119.

Mesoudi, A. (2016). Cultural evolution: A review of theory, findings and controversies. *Evolutionary Biology, 43*(4), 481–497. https://doi.org/10.1007/s11692-015-9320-0.

Mesoudi, A. (2017). Pursuing Darwin's curious parallel: Prospects for a science of cultural evolution. *PNAS, 114*(30), 7853–7860. https://doi.org/10.1073/pnas.1620741114.

Mick, K. M. (2019). *One does not simply preserve Internet memes: Preserving Internet memes via participatory community-based approaches*. Master of Arts, Brandenburg University of Technology. https://doi.org/10.13140/RG.2.2.18093.54240.

Moritz, E. (1990). Memetic science: I - General introduction. *Journal of Ideas, 1*(1), 3–23.

Moscato, P. (1989). On evolution, search, optimization, genetic algorithms and martial arts: Towards memetic algorithms. *Caltech Concurrent Computation Program Report, 826*.

Murray, N., Manrai, A., & Manrai, L. (2014). Memes, memetics and marketing: A state-of-the-art review and a lifecycle model of meme management in advertising. In L. Moutinho, E. Bigné, & A. K. Manrai (Eds.), *The Routledge companion to the future of marketing* (pp. 331–347). London: Routledge.

Neri, F., & Cotta, C. (2012). Memetic algorithms and memetic computing optimization: A literature review. *Swarm and Evolutionary Computation, 2*, 1–14. https://doi.org/10.1016/j.swevo.2011.11.003.

Neri, F., Cotta, C., & Moscato, P. (Eds.). (2012). *Handbook of memetic algorithms*. Berlin: Springer.

Nye, B. D. (2011). Modeling memes: A memetic view of affordance learning. *Publicly accessible Penn Dissertations, 336*. Retrieved from http://repository.upenn.edu/edissertations/336/.

Pagel, M. (2012). *Wired for culture: Origins of the human social mind*. New York: W.W. Norton & Company.

Patzelt, W. J. (Ed.). (2007). *Evolutorischer Institutionalismus: Theorie und exemplarische Studien zu Evolution, Institutionalität und Geschichtlichkeit*. Würzburg: Ergon.

Patzelt, W. J. (Ed.). (2012). *Parlamente und ihre Evolution: Forschungskontext und Fallstudien*. Baden-Baden: Nomos.

Patzelt, W. J. (2015a). Der Schichtenbau der Wirklichkeit im Licht der Memetik. In B. P. Lange & S. Schwarz (Eds.), *Die menschliche Psyche zwischen Natur und Kultur* (pp. 170–181). Lengerich: Pabst.

Patzelt, W. J. (2015b). Was ist "Memetik"? In B. P. Lange & S. Schwarz (Eds.), *Die menschliche Psyche zwischen Natur und Kultur* (pp. 52–61). Lengerich: Pabst.

Patzelt, W. J. (2017). Comparative politics and biology. In S. A. Peterson & A. Somit (Eds.), *Handbook of biology and politics*. Cheltenham: Edward Elgar.

Pech, R. J. (2003). Inhibiting imitative terrorism through memetic engineering. *Journal of Contingencies and Crisis Management, 11*(2), 61–66. https://doi.org/10.1111/1468-5973.1102002.

Phuaphanthong, T. (2014). *A memetic theory of interorganizational information systems (IOIS) emergence and evolution: A longitudinal case study of IOIS for trade facilitation*. Doctoral dissertation, University of Hawai'i at Mānoa.

Popper, K. R. (1972). *Objective knowledge: An evolutionary approach*. Oxford: Oxford University Press.
Popper, K. R. (1974a). Replies to my critics. In P. A. Schilpp (Ed.), *The philosophy of Karl Popper* (pp. 961–1197). Lasalle, IL: Open Court.
Popper, K. R. (1974b). The rationality of scientific revolutions. In R. Harrg (Ed.), *Problems of scientific revolution* (pp. 72–101). Oxford: Clarendon Press.
Popper, K. R. (1978). Natural selection and the emergence of mind. *Dialectica, 32*(3–4), 339–355.
Price, I. (1995). Organizational memetics? Organizational learning as a selection process. *Management Learning, 26*(3), 299–318. https://doi.org/10.1177/1350507695263002.
Prosser, M. B. (2006). *Memetics - A growth industry in US military operations*. Master's thesis, School of Advanced Warfighting, Marine Corps University.
Pyper, H. S. (1998). The selfish text: The Bible and memetics. In J. C. Exum & S. D. Moore (Eds.), *Biblical studies/cultural studies: The third Sheffield colloquium (Gender, Culture, Theory, 7)* (pp. 70–90). Sheffield: Sheffield Academic Press.
Reisman, K. (2013). Cultural evolution. In M. Ruse (Ed.), *The Cambridge encyclopedia of Darwin and evolutionary thought* (pp. 428–435). Cambridge: Cambridge University Press.
Richerson, P. J., & Boyd, R. (2005). *Not by genes alone: How culture transformed human evolution*. Chicago: University of Chicago Press.
Rose, N. (1998). Controversies in meme theory. *Journal of Memetics - Evolutionary Models of Information Transmission, 2*. Retrieved from http://cfpm.org/jom-emit/1998/vol2/rose_n.html.
Roy, D. (2017). Myths about memes. *Journal of Bioeconomics, 19*(3), 281–305. https://doi.org/10.1007/s10818-017-9250-2.
Russ, H. (2014). *Memes and organisational culture: What is the relationship?* Unpublished doctoral dissertation, University of Western Sydney, Sydney.
Salwiczek, L. (2001). Grundzüge der Memtheorie. In W. Wickler & L. Salwiczek (Eds.), *Wie wir die Welt erkennen: Erkenntnisweisen im interdisziplinären Diskurs* (pp. 119–201). Freiburg: Alber.
Schlaile, M. P., & Ehrenberger, M. (2016). Complexity, cultural evolution, and the discovery and creation of (social) entrepreneurial opportunities: Exploring a memetic approach. In E. S. C. Berger & A. Kuckertz (Eds.), *Complexity in entrepreneurship, innovation and technology research: Applications of emergent and neglected methods* (pp. 63–92). Cham: Springer.
Segev, E., Nissenbaum, A., Stolero, N., & Shifman, L. (2015). Families and networks of internet memes: The relationship between cohesiveness, uniqueness, and quiddity concreteness. *Journal of Computer-Mediated Communication, 20*(4), 417–433. https://doi.org/10.1111/jcc4.12120.
Shepherd, J. (2002). *An evolutionary (memetic) perspective on 'how and why does organizational knowledge emerge?'* Doctoral dissertation, Graduate School of Business, University of Strathclyde.
Shepherd, J., & McKelvey, B. (2009). An empirical investigation of organizational memetic variation. *Journal of Bioeconomics, 11*(2), 135–164. https://doi.org/10.1007/s10818-009-9061-1.
Shifman, L. (2014). *Memes in digital culture*. Cambridge, MA: The MIT Press.
Simmel, E. (2009). *Insufficiencies of language: A memetic approach to language-speaker conflicts*. Saarbrücken, Germany: VDM Verlag Dr. Müller.
Sober, E. (1984). *The nature of selection: Evolutionary theory in philosophical focus*. Cambridge, MA: MIT Press.
Sparkes-Vian, C. (2014). *The evolution of propaganda: Investigating online electioneering in the UK general election of 2010*. De Montford University, Leicester. Retrieved from http://hdl.handle.net/2086/10752.
Speel, H.-C. (1998). Memes are also interactors. In J. Ramaekers (Ed.), *Proceedings of the 15th international congress on cybernetics* (pp. 402–407). Association Internationale de Cybernétique.
Speel, H.-C. (1999). Memetics: On a conceptual framework for cultural evolution. In F. Heylighen, J. Bollen, & A. Riegler (Eds.), *The evolution of complexity: The violet book of "Einstein meets Magritte"* (pp. 229–254). Dordrecht: Kluwer Academic Publishers.
Sperber, D. (1996). *Explaining culture: A naturalistic approach*. Oxford: Blackwell.

Spitzberg, B. H. (2014). Toward a model of meme diffusion (M^3D). *Communication Theory, 24*(3), 311–339. https://doi.org/10.1111/comt.12042.

Stake, J. E. (2001). Are we buyers or hosts? A memetic approach to the First Amendment. *Alabama Law Review, 52*(4), 1213–1268.

Stanovich, K. E. (2005). *The robot's rebellion: Finding meaning in the age of Darwin*. Chicago: The University of Chicago Press.

Stewart-Williams, S. (2018). *The ape that understood the universe: How the mind and culture evolve*. Cambridge: Cambridge University Press.

Sukopp, T. (2010). Rätselhafte evolutionäre Ethik? Was sind und wie "wirken" Meme? *Erwägen Wissen Ethik, 21*(2), 301–304.

Taillard, M., & Giscoppa, H. (2013). *Psychology and modern warfare: Idea management in conflict and competition*. New York: Palgrave Macmillan.

Taylor, J. R., & Giroux, H. (2005). The role of language in self-organizing. In G. A. Barnett & R. Houston (Eds.), *Advances in self-organizing systems* (pp. 131–167). Cresskill, NJ: Hampton Press.

Velikovsky, J. T. (2016). The holon/parton theory of the unit of culture (or the meme, and narreme): In science, media, entertainment, and the arts. In A. M. Connor & S. Marks (Eds.), *Creative technologies for multidisciplinary applications* (pp. 208–246). Hershey, PA: IGI Global.

Velikovsky, J. T. (2018). The holon/parton structure of the meme, or the unit of culture. In M. Khosrow-Pour (Ed.), *Encyclopedia of information science and technology* (pp. 4666–4678). https://doi.org/10.4018/978-1-5225-2255-3.ch405.

von Bülow, C. (2013a). Mem. In J. Mittelstraß (Ed.), *Enzyklopädie Philosophie und Wissenschaftstheorie* (2nd ed., Vol. 5, pp. 318–324). Stuttgart: Metzler.

von Bülow, C. (2013b). Meme. [English translation of the (German) article "Mem". In J. Mittelstraß (Ed.), *Enzyklopädie Philosophie und Wissenschaftstheorie* (2nd ed., Vol. 5, pp. 318–324). Stuttgart: Metzler]. Retrieved from https://www.philosophie.uni-konstanz.de/typo3temp/secure_downloads/87495/0/de0f56268a8ad66b13cfc7652e092ce47ea79fb6/meme.pdf.

von Sydow, M. (2012). *From Darwinian metaphysics towards understanding the evolution of evolutionary mechanisms: A historical and philosophical analysis of gene-Darwinism and universal Darwinism*. Göttingen: Universitätsverlag Göttingen.

Weeks, J., & Galunic, C. (2003). A theory of the cultural evolution of the firm: The intraorganizational ecology of memes. *Organization Studies, 24*(8), 1309–1352. https://doi.org/10.1177/01708406030248005.

Wegener, F. (2015). *Memetik. Der Krieg des neuen Replikators gegen den Menschen* (3rd ed.). Gladbeck: Kulturförderverein Ruhrgebiet e.V.

Weng, L. (2014). *Information diffusion on online social networks*. Doctoral dissertation, School of Informatics and Computing, Indiana University. Retrieved from http://lilianweng.github.io/papers/weng-thesis-single.pdf.

Weng, L., Flammini, A., Vespignani, A., & Menczer, F. (2012). Competition among memes in a world with limited attention. *Scientific Reports, 2*, 335. https://doi.org/10.1038/srep00335.

Westoby, A. (1994). *The ecology of intentions: How to make memes and influence people: Culturology*. Draft-Working Paper, Center for Cognitive Studies, Tufts University. Retrieved from http://ase.tufts.edu/cogstud/dennett/papers/ecointen.htm.

Wiggins, B. E., & Bowers, G. B. (2015). Memes as genre: A structurational analysis of the memescape. *New Media & Society, 17*(11), 1886–1906. https://doi.org/10.1177/1461444814535194.

Wilkins, J. S., & Bourrat, P. (2019). Replication and reproduction. In E. N. Zalta (Ed.), *The Stanford encyclopedia of philosophy, Summer 2019 edition*. Retrieved from https://plato.stanford.edu/archives/sum2019/entries/replication/.

Williams, G. C. (1986). Comments by George C. Williams on Sober's The Nature of Selection. *Biology & Philosophy, 1*(1), 114–124. https://doi.org/10.1007/BF00127091.

Williams, P. A. (2002). Of replicators and selectors. *The Quarterly Review of Biology, 77*(3), 302–306. https://doi.org/10.1086/341995.

Williams, R. (2000). The business of memes: Memetic possibilities for marketing and management. *Management Decision, 38*(4), 272–279. https://doi.org/10.1108/00251740010371748.

Williams, R. (2002). Memetics: A new paradigm for understanding customer behaviour? *Marketing Intelligence & Planning, 20*(3), 162–167. https://doi.org/10.1108/02634500210428012.

Williams, R. (2004). Management fashions and fads: Understanding the role of consultants and managers in the evolution of ideas. *Management Decision, 42*(6), 769–780. https://doi.org/10.1108/00251740410542339.

Worden, R. P. (2004). Words, memes and language evolution. In C. Knight, J. R. Hurford, & M. Studdert-Kennedy (Eds.), *The evolutionary emergence of language* (pp. 353–371). Cambridge: Cambridge University Press.

Ziman, J. (Ed.). (2000). *Technological innovation as an evolutionary process*. Cambridge: Cambridge University Press.

Chapter 3
A Case for Econememetics? Why Evolutionary Economists Should Re-evaluate the (F)utility of Memetics

Michael P. Schlaile

Abstract The paper presented in this chapter contributes to evolutionary approaches in economics and related disciplines by discussing the potential of a memetic perspective. The central aim of this endeavor is to reveal and establish connections between various rather fragmented lines of research. The point of departure is the observation that both imitation and cultural evolution have not received sufficient attention from evolutionary economists. Building on a review of criticisms and definitions of both memes and cognate entities in evolutionary economics, an "informationalist" perspective is proposed that is also in line with the notion of complex population systems. Moreover, by shedding light on similarities and implications of both memetics and the rule-based approach to evolutionary economics, we are able to create links to imitation heuristics and evolutionary institutionalism. In summary, the chapter lays out four propositions that can be used as starting points for further work at the frontiers of memetics and evolutionary social science.

3.1 Introduction

Even long before the previous economic and financial crises that started about a decade ago, the rational, self-interested *homo (o)economicus* seemed to have become an increasingly "endangered species" (Persky 1995). The axioms and assumptions of mainstream economics have been vigorously criticized (see also Schlaile et al.

This paper is single authored and hitherto unpublished. I have benefited from presenting earlier versions at the 16th International Schumpeter Society Conference 2016 in Montreal, at the Generalized Theory of Evolution Conference 2018 in Düsseldorf, and at the European Academy of Management Conference 2018 in Reykjavik. I would like to thank Daniel Dennett, Kurt Dopfer, Richard Nelson, Gerhard Schurz, Michael Schramm, and Andreas Pyka for helpful comments, suggestions, and encouragement. I am also indebted to the valuable comments from two anonymous reviewers on an earlier draft submitted to (but ultimately rejected by) the *Journal of Evolutionary Economics*.

M. P. Schlaile (✉)
Institute of Economics (520), Institute of Education, Labor and Society (560), University of Hohenheim, Wollgrasweg 23, 70599 Stuttgart, Germany
e-mail: schlaile@uni-hohenheim.de

© The Author(s), under exclusive license to Springer Nature Switzerland AG 2021
M. P. Schlaile (ed.), *Memetics and Evolutionary Economics*, Economic Complexity and Evolution, https://doi.org/10.1007/978-3-030-59955-3_3

2018a), giving rise to a vast range of *heterodox* schools of economic thought (e.g., Keen 2011; Lawson 2006; Lee 2009). This conceptual chapter contributes to a particular heterodox economic school of thought, namely, to *evolutionary economics* (for overviews, see, e.g., Dopfer 2016; Hodgson 2011; Hodgson and Lamberg 2018; Nelson et al. 2018; Stoelhorst 2014; Witt 2014, 2016; Winter 2017).[1] In this chapter, I discuss the potential of a memetic approach to economic evolution with the aim of revealing links between different but complementary concepts and approaches in evolutionary economics and related disciplines. This endeavor is motivated by two—at a first glance seemingly unrelated—starting points.

The first starting point is the importance of *imitation* (not just) for economic systems and innovation processes (see, e.g., Alchian 1950; Iwai 1984; Shenkar 2010a, 2010b; Tarde 1890, 1903). As also Blaine Fowers writes: "In recent years, imitation has become recognized as central to defining features of humans such as intelligence, language, and cumulative culture" (Fowers 2015, p. 131; see also Dugatkin 2000). Despite this importance, it has been purported that imitation is underresearched by both economists and management scholars (e.g., Niosi 2012, 2017).[2]

The second starting point is the underrepresentation of *cultural evolution* in (evolutionary) economic analysis. In Yuichi Shionoya's words:

> Recent works on evolutionary economics, sometimes labeled 'neo-Schumpeterian economics', are largely confined to the studies of economic development and technological change. ... Compared with Schumpeter's original view of *sociocultural* development, the current conception of evolution is narrow ... (Shionoya 2008, p. 15, emphasis added).

Arguably, various economists, especially those intellectually rooted in the (neoclassical) "mainstream," still ignore the economic significance of culture or just take it as a given phenomenon outside their explanatory system, a position Eric Jones (1995) has termed *cultural nullity*. Although there are exceptions (e.g., see Becker 1996; Gorodnichenko 2011; Guiso et al. 2006; Klasing 2013; Spolaore and Wacziarg 2013), the position of cultural nullity has long appeared to be the "conventional wisdom of the dominant group" (coined by Waddington 1977)[3] among mainstream economists, but it should go without saying that I will not assume such a position here. The second point of view identified by Jones is *cultural fixity*, which assumes that "cultures should essentially be taken as givens to which the economy adapts" (Jones 1995, p. 276). The third and last approach has been termed *cultural reciprocity*, which neither considers culture as given nor the economy as divorced from the cultural context, hence, culture—seen as a process and not an end-state—may also be influenced by economic processes and vice versa (Jones 1995; see also Nau

---

[1] For a possible taxonomy of evolutionary economic contributions see, e.g., Robert et al. (2017).

[2] Interestingly, Joseph Schumpeter, who is sometimes even hailed as a "prophet of innovation" (McCraw 2007), already mentioned "that needs and their visible satisfaction immediately lead to a *contagious effect* on the economic agents in the vicinity" (Schumpeter 2002 [1912], p. 104 [483], emphasis added).

[3] This expression was abbreviated to "COWDUNG" by Waddington (1977, p. 16), which may be appropriate also in this context. See also Galbraith (1998) for an explanation of the concept of *conventional wisdom*.

2004). While this third position seems to be the most adequate perspective for this chapter, one may be tempted to use the notion of *coevolution* instead of reciprocity (see also Almudi and Fatas-Villafranca 2018, on a related note).

Although there are indeed various scholars interested in the interplay of culture and economy (e.g., see Herrmann-Pillath 2010a; Schramm 2008; Throsby 2001; or Beugelsdijk and Maseland 2011, for an overview), the realization that culture *evolves* according to processes very similar to biological evolution seems to have been rather neglected by many economists and management scholars either out of ignorance or deliberately for methodological (or ideological) reasons.[4] This "avoidance strategy" is simply inappropriate against the backdrop of various important advances in research on cultural evolution.[5] Whereas most of the studies on cultural evolution also appear to agree that the cultural evolutionary process follows broadly Darwinian principles, it is still open to debate how far the analogies to biological evolution should be pushed (e.g., Acerbi and Mesoudi 2015; Claidière et al. 2014; Creanza et al. 2017; Lewens 2015, 2019; Smith et al. 2018). One of the approaches to cultural evolution is *memetics* (or *meme theory*), which also heavily relies on Darwinian foundations and biological analogies or metaphors. As Susan Blackmore puts it:

> Memetics is a theory of cultural evolution based on the idea that behaviors, skills, habits, stories, and technologies that are copied from person to person in culture act as a second replicator. That is, they are information that is copied with variation and selection, and they therefore sustain a new evolutionary process, both cooperating and competing with the old (Blackmore 2010, p. 255).

Why is this relevant and what does this mean for our two starting points? The pivotal thesis is that by taking memetics more seriously, (evolutionary) economists may tackle both of these issues (of understudied imitation and the often neglected cultural evolution) simultaneously and build bridges between some of the rather fragmented approaches in evolutionary economics.[6] For general overviews and some introductory remarks on the memetic approach, the reader may refer to Barrett et al. (2002, Chap. 13), Buskes (2013), Cartwright (2008, Chap. 16), Heylighen and Chielens (2009), Jesiek (2003), Roy (2017), and von Bülow (2013).[7] For a summary of arguments in favor of memetics, see also Blackmore (2003, 2010), Gers (2008), and Dennett (2013, Chap. 52; 2017).

In order to understand how memetics may provide a way to address the above issues, we have to start with the original definition by Richard Dawkins (1976).

---

[4]Of course, this is not to say that cultural change has been neglected in these disciplines. The word *evolve* should thus be treated with care and needs further clarification, as, for example, also Hodgson (2000, 2002) and Dollimore (2006) stress.

[5]For an overview, see, for example, Boyd and Richerson (1985, 2005), Buskes (2013, 2015, 2016), Carroll et al. (2017), Creanza et al. (2017), Distin (2011), Mesoudi (2007, 2011, 2016a, 2016b, 2017), Mesoudi et al. (2006), Richerson and Christiansen (2013), Whiten et al. (2012).

[6]On the 'fragmentation' of evolutionary economics, see, e.g., Hodgson and Lamberg (2018), Stoelhorst (2014), or Witt (2014, 2016).

[7]For an easily accessible introduction to memetics, see also http://www.practicalmemetics.com/Concepts/Memetics101/Memetics101/intro.html.

Dawkins introduced the *meme* as a "unit of cultural transmission, or a unit of *imitation*" (Dawkins 2016, p. 249, italics in original) in the first edition of *The Selfish Gene*, where he also referred to the meme as a new *replicator* (Dawkins 1976; see also Ball 1984; Dennett 2002; Godfrey-Smith 2000; Wilkins 1998; Wilkins and Hull 2014). Blackmore recently clarified again, with reference to Dawkins (1976) and Dennett (2006):

> The difference between memetics and other theories of cultural evolution is that for memetics cultural elements (memes) are replicators ... and therefore, like genes, have replicator power. When memes compete for survival, they do so not primarily for the benefit of the genes of their carriers but for their own benefit (Blackmore 2016, p. 22).

As Dawkins (2015) himself noted not so long ago, significant advances in memetics have been made by others, including Robert Aunger (2002), Blackmore (e.g., 1999), and Daniel Dennett (e.g., 1990, 1995, 2002, 2017). Some of the arguments against memetics that can be found in the literature (e.g., in Aunger 2000; Frank 2010; Hannon and Lewens 2014; Ingold 2013; Lewens 2007, 2015, 2019; Polichak 1998) may be justified, some may not (see also Chap. 11 in Dennett 2017, for responses to some of these objections); but it seems that several economists rejecting the notion of memes do so mostly on the basis of a critique of (what they regard as) Dawkins' (1976) notion, thereby neglecting other—and more recent—discussions and successful scientific applications of the memetic perspective to other disciplines. For example, various economists are indeed sympathetic to the notion of replicators and the concept of a *universal* or *generalized Darwinism* (e.g., Aldrich et al. 2008), but many of them reject or avoid the notion of memes and seem to neglect the more recent literature on memetics (see Chap. 2 above), which creates the impression that they may be throwing the proverbial baby out with the bathwater.[8] This leads us to **general proposition I:** *The (f)utility of integrating memetics with evolutionary economics should be re-evaluated.*

Hence, the central aim of this chapter is to fathom the potential of including elements of memetics into frameworks of evolutionary economics. At this point, however, it is important to remember that there is also the general debate about the merits and limits of generalized Darwinism for the study of sociocultural and

---

[8] Readers who have followed the respective discussions in the *Journal of Evolutionary Economics*, for example, may argue that this seems like an overly casual dismissal of important arguments, especially given that Aldrich et al. (2008) propose alternatives such as habits, routines, customs, etc. However, we will come back to some of these alternatives below. Moreover, it is important to note that there are also various economists embracing some version of memetics, including many of the contributors in Ziman (2000), Ken Binmore (e.g., 1998), who provides his own definition of memes based on Dawkins (1976), Carsten Herrmann-Pillath (2000, 2010b, 2013), who prefers to follow Aunger's (2002) notion of *neuromemes*, and most recently also Nobel laureate Robert Shiller (2017). For Binmore's definition of memes and for Aunger's definition of neuromemes see Appendix: Selected Examples of Definitions or Uses of Memes.

economic evolution. Although an explicit contribution to this debate is not among the aims of this chapter, it touches upon selected issues.[9]

In an earlier discussion, I proposed the term *economemetics* to describe this highly interdisciplinary research endeavor that aims to explore the potential of utilizing memetic approaches for (evolutionary) economics (Schlaile 2013).[10] I will use the term economemetics in this sense also for the purpose of this chapter.

The paper presented in this chapter is structured as follows: Sect. 3.2 reviews relevant literature and provides general clarifications and classifications necessary for the subsequent sections. The implications of an "informationalist" view on memes are discussed in Sect. 3.3, where I suggest a complex systems perspective on memes combined with the economics of attention. In Sect. 3.4, I illuminate similarities and implications of the so-called rule-based approach also with an eye to evolutionary institutionalism. I conclude in Sect. 3.5 by illustrating how several strands of literature may be connected by economemetics, summarizing this paper's key contributions, and giving an outlook on future research potentials.

3.2 Some Clarification and Classification

First of all, what is culture? Before we can contribute anything to discussions about cultural evolution, we should remember that there are myriads of definitions of the term culture itself (e.g., see Baldwin et al. 2006, for a collection of over three hundred from across disciplines). So, where do we start? The good news is that Lee Cronk provides a helpful suggestion: "Any discussion of the concept of culture must begin with Edward Burnett Tylor, the Englishman who gave the term its first technical definition" (Cronk 1999, p. 3). Tylor defined culture as "that complex whole which includes knowledge, belief, art, morals, law, custom, and any other capabilities and habits acquired by man as a member of society" (Tylor 1871, p. 1).[11] While this definition may be broad enough to be accepted by almost anyone, it is also somewhat impractical and already gives a clue to why there are so many different definitions

---

[9]Readers interested in this debate in full detail can, for example, refer to Aldrich et al. (2008), Bagg (2017), Blute (2010), Breslin (2010, 2011), Buenstorf (2006), Callebaut (2011a, 2011b), Cordes (2006, 2007a, 2007b), Dawkins, (2017), Dollimore (2006, 2014), Essletzbichler and Rigby (2007), Geisendorf (2009), Hodgson (2002, 2005, 2007a, 2013b, 2013a), Hodgson and Knudsen (2006, 2008, 2010, 2011, 2012), Jagers op Akkerhuis et al. (2016), Müller (2010), Nelson (2006, 2007), Pelikan (2011), Reydon (2016), Reydon and Scholz (2015), Scholz and Reydon (2013), Schubert (2014), Schurz (2011), Stoelhorst (2008a, 2008b), Tang (2017), Thomas (2018), von Sydow (2012), Wäckerle (2014), and Wortmann (2010).

[10]Note that there is also the framework of "MEMEnomics" that has been developed by Said Dawlabani (2013) based on the vMEMEs approach by Beck and Cowan (1996). According to the project's website, memenomics aims to examine "the long-term effects of economic policy on culture" (http://www.memenomics.com/what-is-memenomics).

[11]As Alex Mesoudi puts it: "Ignoring the Victorian sexism in restricting culture to the male half of the species and the anthropocentrism in restricting culture to humans, the key phrase here is 'acquired as a member of society,' that is, socially transmitted" (Mesoudi 2011, p. 221).

of memes (see Appendix: Selected Examples of Definitions or Uses of Memes for examples).[12] Nevertheless, Tylor's definition contains an important cue, which I will come back to later (in Sect. 3.3.1), namely, *complexity*.

For the purpose of this paper, we may narrow it down a bit and adopt the definition proposed by Robert Boyd and Peter Richerson:[13]

> *Culture is information capable of affecting individuals' behavior that they acquire from other members of their species by teaching, imitation, and other forms of social transmission* (Boyd and Richerson 2005, p. 6, italics in original).

The next step is to acknowledge that there is a level of societal evolution that is crucial from the viewpoint of memetics. Namely, as Peter Godfrey-Smith explains:

> The evolution of human behavior features an important role for copying and imitation, and other forms of social learning. *When behaviors and ideas spread by copying, this raises the possibility of an evolutionary dynamic in the pool of ideas and behaviors themselves* (Godfrey-Smith 2014, p. 136, emphasis added).

The importance of imitation has also been affirmed by researchers studying adaptive behavior and cognition,[14] who identified several *heuristics*, including *imitate-the-successful* (also called *imitate-the-best*)[15] and *imitate-the-majority* or *imitate-your-peers* (e.g., see Gigerenzer et al. 2011, with reference to Boyd and Richerson 2005; see also Chap. 8.3 in Godfrey-Smith 2009, on a related note). Already at the end of the nineteenth century, the French sociologist Gabriel Tarde published *The Laws of Imitation* (Tarde 1890, 1903). Tarde is a central figure in two ways: First, he can be considered to be one of the important "forefathers" of memetics (e.g., Bammé 2009; Marsden 2000, P. Mitchell 2012; see also Schmid 2004, on a related but critical note). Second, he also seems to have influenced or foreshadowed some important elements of Schumpeter's works (e.g., Barry and Thrift 2007; Kobayashi 2014, 2015a, 2015b; Taymans 1950) as well as diffusion research (Katz 2006; Kinnunen 1996; Rogers 2003). Against this background, it is quite understandable that Bruno Latour (2002, p. 131) has criticized Blackmore (1999) for not having mentioned Tarde despite her emphasis on imitation.

Once we take a closer look at the different definitions of memes compiled in Appendix: Selected Examples of Definitions or Uses of Memes, we can see that there are (at least) two important levels to classify the approaches (see also Baraghith 2015; Schurz 2011, Chap. 9.4, for related discussions): First, there are the definitions that seek to identify memes with neural substrates or confine them to human brains

---

[12] Finkelstein (2008) and Finkelstein and Ayyub (2010) also present a list of various definitions of the meme. However, unfortunately, they do not mention the sources.

[13] We could also take up the definition proposed by Mesoudi (2011, p. 2), which is very similar to Boyd and Richerson's but omits the influence on behavior.

[14] Note, however, that a critique of imitation and arguments against its importance in the context of the evolution of traditions have been brought forward by Olivier Morin (2016) and, on a related note, by Pascal Boyer (2018, pp. 250ff.).

[15] In Dawkins and Wong's words: "People are most apt to copy their memes from admired models" (Dawkins and Wong 2016, p. 326). Note that this heuristic is closely related to so-called "prestige bias" in cultural transmission (e.g., Henrich and Gil-White 2001; Mesoudi 2011, 2017).

(e.g., Dawkins' later definition in *The Extended Phenotype*, or the ones by Aunger, Delius, and Wilson) versus the definitions that regard memes as (substrate-neutral) informational entities (e.g., Dennett, Esser, Kappelhoff). We could call this first level "neuralist vs. informationalist". Second, according to Marion Blute (2010), there was also "a fratricidal war between adherents of the gene-like biologically adaptive view ... and adherents of the virus-like biologically maladaptive view" (Blute 2010, p. 113; see also Blute 2005). The latter point of view may be associated with notions such as "viruses of the mind" (e.g., Brodie 1996; Dawkins 1993), "thought contagion" (Lynch 1996), or "contagious ideas" (Cullen 2000). Notably, this virus-like position has also been advocated by Malcolm Gladwell in his bestseller *The Tipping Point*, albeit without explicit reference to memes:

> [T]he best way to understand the emergence of fashion trends, the ebb and flow of crime waves, or, for that matter, the transformation of unknown books into bestsellers, or the rise of teenage smoking, or the phenomenon of word of mouth, or any number of the other mysterious changes that mark everyday life is to think of them as epidemics. Ideas and products and messages and behaviors spread just like viruses do (Gladwell 2000, p. 7).

Due to this paper's focus on imitation and Boyd and Richerson's definition of culture adopted above, I will not confine myself to memes actualized solely as neural substrates. Instead, the definitions making no assumption about the physical substrate can be assumed to be better suited for bridging fragmented approaches across disciplines, and I thus follow Maarten Boudry (2018b), Dennett (e.g., 2017), and other "informationalists" in this respect. With an eye to the second differentiation, however, I follow Blute (2005, 2010) and treat both the gene-like and the virus-like views as valid as long as the analogies are not taken too literally. In Cronk's words: "Memes are in some ways similar to genes, but they are also sometimes similar to viruses. Like viruses and other pathogens, memes are passed from person to person and may not necessarily be helpful to the people who catch them" (Cronk 1999, p. xii).[16]

As Edward O. Wilson explained already two decades ago, the

> notion of a cultural unit ... has been around for over thirty years, and has been dubbed by different authors variously as mnemotype, idea, idene, meme, sociogene, concept, culturgen, and culture type. The one label that has caught on the most, and for which I now vote to be the winner, is meme (Wilson 1998, p. 148).

In a similar manner, evolutionary economists have also proposed various candidate units (or even replicators) in an economic context, including the (organizational) *routines* of Richard Nelson and Sidney Winter (1982),[17] the *habits* of Geoffrey Hodgson and Thorbjørn Knudsen (2010), Eric Beinhocker's (2006) *modules*, and the *rules* of

---

[16]In fact, recent works on information diffusion and social networks suggest that only some memes may "go viral" and spread like *simple contagions* (such as infectious diseases) (Weng et al. 2014), whereas others spread like *complex contagions*, which are sensitive to *social reinforcement* and *homophily* (Weng 2014; Weng et al. 2013; see also Lerman 2016, for a review).

[17]For reviews and further clarifications, see also Becker and Knudsen (2012), Breslin (2016), Hodgson (2009), and Knudsen (2008).

Kurt Dopfer and collaborators.[18] While it would go well beyond the scope of this paper to discuss all of these alternatives in detail (see also Breslin 2016, on a related discussion), I will at this juncture briefly address some critiques by prominent evolutionary economists. Almost two decades ago, Nelson (2001) himself noted: "The concept of routines is analytically similar to the genes in biological theory, or the memes or culturgenes [sic] in sociobiology" (Nelson 2001, p. 170). More recently, however, Nelson put this statement into perspective in a personal communication about memes at the 16th International Schumpeter Society conference: "I don't believe that the lion's share of practices or ideas that drive practices can be described as simply as genes" (R.R. Nelson, personal communication in Montreal, 7 July 2016). In a similar vein, Ulrich Witt already wrote two decades ago that "ideas and images are not only well known for their subjectivity and volatility, they are also created and propagated in a way that seems to follow its own regularities, and these cannot be compared to the regularities governing the genetic mechanisms of recombination and inheritance" (Witt 1999, p. 23). These arguments may, however, be countered by stressing that memetics is not about stubbornly sticking to close analogies between genes and memes (see also Runciman 2015, p. 198, in this regard). Moreover, as also Hodgson (2013a, p. 985) writes, an "adoption of Darwinian principles is not primarily a matter of analogy, but of ontological communality at an abstract level." However, in *Darwin's Conjecture*, Hodgson and Knudsen (2010, especially Chap. 6.6) raise several issues with memes, and so do various other critics. Discussing all of them would require another much more lengthy treatment (which can actually be found in Dennett 2017), so I will only pick up some selected arguments here, which have not received much attention in other rebuttal attempts. For example, Hodgson and Knudsen argue:

> In the memetics literature, the nature of ideas and the causal mechanisms by which ideas lead to phenotypic behavior are rarely spelled out. As a result, in a very real sense, memetics is insufficiently Darwinian: it does not identify the detailed, causal mechanisms involved (Hodgson and Knudsen 2010, p. 132).

One can counter this by stating that, first, Ted Cloak, whose *cultural instructions* (Cloak 1975) are an important progenitor of memes, has recently proposed a way to address this issue (Cloak 2015), and second, as Stephen Shennan writes

> Even if the meme concept in the strict sense is problematical, the word meme has been such a successful meme itself that it represents a useful shorthand way of referring to the idea that culture is an evolutionary system involving inheritance ... We can ask what are the population level processes characteristic of this inheritance system. This is what biologists did before they understood genetics. They could still measure the heritability of particular traits from one generation to the next without knowing the mechanisms involved (Shennan 2002, p. 48).

John Wilkins and David Hull (2014) also clarify, albeit in a slightly different context:

---

[18] I will come back to rules in Sect. 3.4.

> To be sure, when we make copies on a Xerox machine, we are interested in the texts, figures or just scrawls that appear on these sheets of paper. We are not interested in how the Xerox machine works, even if it does *all* the work ... Numerous mechanisms exist for passing on information from one replicator to another. Sometimes the material in which the information is encoded gets passed on; sometimes not ... [I]f replication were simply the passing on of structure largely intact without any subsequent translation, we would not be tempted to employ terms such as 'information' in connection with it (Wilkins and Hull 2014, no pagination, italics in original).

Another one of Hodgson and Knudsen's critiques essentially concerns issues related to the "locus" of the meme:

> The meme as behavior cannot serve as a replicator because behavior is an outcome rather than a mechanism or a disposition. Furthermore, the concept of meme as idea faces the problem of the identification of its material substrate and of the underlying emotions or dispositions on which the ideas are grounded. The meme as idea points to an untenable dualism of the ideal and the material worlds (Hodgson and Knudsen 2010, p. 134–135).

Some aspects of this problem have already been addressed by Blackmore's (1999) distinction between *copying-the-instruction* and *copying-the-product*. However, this discussion also tends to lead to the central questions of *philosophy of mind* and debates about *monism* versus *dualism*, *materialism* versus *idealism*, and, essentially, *the mind–body problem*, which can obviously not be adequately tackled within this paper. However, various related issues (including ideal/material and agency/structure "dualisms") have also been addressed by Blute (2010). In this regard, it may be helpful to take up the classification recently proposed by Michael Schlaile and Marcus Ehrenberger (2016). Our "three-dimensional" (p–i–e) view on memes draws inspiration from Karl Popper's (1972, 1974) *evolutionary epistemology* and his *three worlds*, among others (e.g., the three levels of memes proposed by Franz Wegener 2015). This approach is in line with other scholars that also advocate a Popperian view on memes (e.g., Chesterman 2016, Patzelt 2015b).

The p–i–e trichotomy (see Fig. 3.1) can be summarized as follows: We may differentiate between *p-memes* (primal memes), *i-memes* (mental representations), and *e-memes* (environmental representations). Whereas p-memes can be seen as genuine "bits" of information (or objects of Popper's world 3),[19] i-memes are individual mental representations of memes, for example, as elements of cultural *schemata* that help transform information into knowledge and serve as models *of* the world as well as models *for* the world (see also Schlaile and Ehrenberger 2016, pp. 68–70). These i-memes may also be linked to Cloak's (1975) *i-culture* and his recent neurological model of the meme (Cloak 2015), and perhaps even to the "neuralist" views of authors like Aunger (2002), Delius (1991), or Wilson (1998). E-memes can in turn be regarded as the environmental or artifactual representation (or "reification")

---

[19]Especially in the sense that these "world 3 memes" may not necessarily possess a world 3.1 materialization as described by Popper (1974, p. 1051), or in the sense that nobody currently recognizes them as such, or they are only *potentially* replicable representational contents, see Patzelt (2015b, p. 57–58). In Beinhocker's words, "while all physically rendered designs are actualizations of ideas, it does not follow that all ideas or possible schemata are or can be actualized" (Beinhocker 2011, p. 402). In this sense, p-memes may be regarded as the constraints of the opportunity space for a particular culture.

Fig. 3.1 The three-dimensional (P-I-E) perspective on memes. Adapted from Schlaile and Ehrenberger (2016)

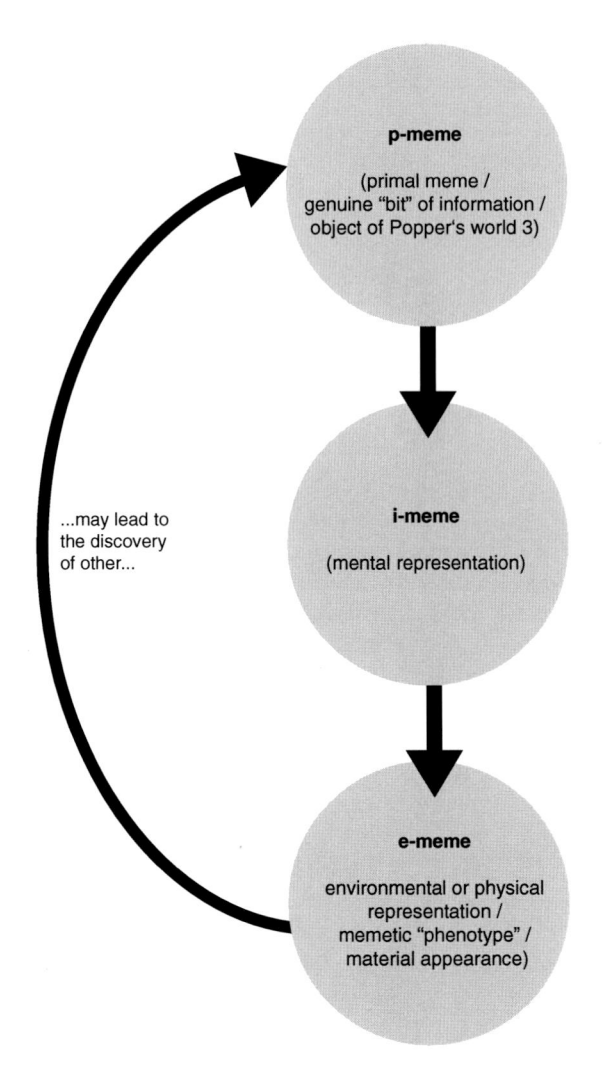

p-meme
(primal meme / genuine "bit" of information / object of Popper's world 3)

...may lead to the discovery of other...

i-meme
(mental representation)

e-meme
environmental or physical representation / memetic "phenotype" / material appearance)

of memes, for example, as (codified in) tools, books, imitable behavior, practices, and the like (see also a similar differentiation between i-memes and e-memes in McNamara 2011).[20]

As a potential solution to their issues with memes, Hodgson and Knudsen (2010) advocate their notion of habits instead of memes as social replicators.[21] However,

---

[20] Note that these p–i–e dimensions may also be seen as different poles of a continuum (see also Fuchs 2001, on a related note) and the suggestion deliberately avoids a rigid replicator-interactor distinction (e.g., Hull 1982, 1988a, 1988b) as this has also caused some debate (e.g., Speel 1998).

[21] According to Hodgson and Knudsen, a "habit is a disposition to engage in previously adopted or acquired behavior that is triggered by an appropriate stimulus. Habits are formed through the repetition of behavior or thought. They are influenced by prior activity and are the basis of both reflective and nonreflective behavior" (Hodgson and Knudsen 2010, p. 137).

at the same time they state that "[h]abit replication also often relies on imitation" (Hodgson and Knudsen 2010, p. 138), which would make the habit (if it is considered a replicator) a meme—at least "by definition" in the sense of those authors focusing on memes as the units of imitation.

Habits have a long history,[22] also in economics (e.g., Veblen 1899), and they have even been conceived as potentially maladaptive. As Jack Vromen writes: "Like Veblen, Marshall held that *many habits survive which are in themselves of no advantage to the human race. Yet, on the whole, Marshall seemed to be convinced that those races survive in which the best habits are developed*" (Vromen 1995, p. 2, emphasis added). Hence, it would seem inappropriate to just equate habits with memes (clearly, not all habits have been acquired via social learning),[23] but I can suggest **proposition 1:** *Not all habits are memes but habits that are replicated via imitation should be treated as a special subset (or subtype) of memes that are (evolutionarily) stabilized especially by repetitive means.*

The final criticism addressed here stems from Martin Mahner and Mario Bunge:

> [T]he notion of a *meme* ... is nothing but a metaphor. Indeed, whereas pieces of DNA (i.e., genes) are actually passed on in the production of offspring, there are no such things as pieces of knowledge (i.e., memes or ideas) that are literally transmitted to other brains (Mahner and Bunge 1997, p. 65, italics in original).

As already noted above, the specific processes involved in cultural transmission are often more complex than suggested by the notion of replicators or imitation (again, see also Wilkins and Hull 2014),[24] but nonetheless, memes do indeed provide a useful metaphor (see also Roy 2017), which brings us to the first part of Mahner and Bunge's critique. We could counter their "nothing but a metaphor" with George Lakoff and Mark Johnson: "The essence of metaphor is understanding and experiencing one kind of thing in terms of another" (Lakoff and Johnson 1980, p. 5, emphasis removed). In the case of memes, the "essence" includes (but is not limited to) an understanding that ideas, habits, rules, and other (cultural) units of imitation can have differential "fitness" as informational entities[25] and are subject to an evolutionary process that

---

[22] For a history of the philosophy of habit see the contributions in Sparrow and Hutchinson (2013).

[23] For example, note also the distinction between "habits of action" and "habits of thought" (e.g., Murdock 1960). Moreover, as Murdock (1965, p. 125) notes: "So long as an innovation ... is practiced by the innovator alone in his society, it is an individual habit and not an element of culture. To become the latter it must be accepted by others; it must be socially shared."

[24] Moreover, as also Dennett (2017, p. 237) writes: "Not all processes of change in culture exhibit features analogous to competition of alleles at loci, but that feature is in any case just one of the dimensions along which evolutionary processes can vary, some more 'Darwinian' than others."

[25] Memetic fitness can take various shapes, as elucidated by von Bülow (2013, p. 4): "*ceteris paribus*, those memes will be fitter (1) that draw more attention to themselves ..., (2) that are psychologically attractive ..., (3) that tend to stick well in memory ... and (4) that strongly stir their bearers into action ..."

involves *variation, selection,* and *retention* (see also Aldrich et al. 2008; Campbell 1960, 1965, 1974, on the importance of this Darwinian triad).[26] Therefore, memes may spread when conditions allow them to do so which need not necessarily coincide with them being helpful to their carriers (i.e., the so-called *meme's eye view*; for a detailed version of this argument, see also Runciman 2002, or Dennett 2011, 2017, Chap. 10). This alternative view is not merely a "fancy way of talking" but rather of *thinking* in a certain way that may help to generate insights that have been overlooked by the "conventional," that is, agent-centered way (Boudry 2018a, 2018b; Boudry and Hofhuis 2018; Dennett 2017, 2018). In a similar vein, Lakoff and Johnson argue "that metaphor is not just a matter of language, that is, of mere words. ... [O]n the contrary, human *thought processes* are largely metaphorical" (Lakoff and Johnson 1980, p. 6, italics in original). However, with regard to the meme's eye view and the epidemiological analogies mentioned above, it seems important at this point to repeat Blackmore's (2005) clarification

> that no meme theorist is likely to reject human goals as irrelevant to memetic evolution. The interesting question is what role they play. Are human goals the ultimate design force for culture ... or are they just one of many factors in a Darwinian design process acting on memes ...? (Blackmore 2005, p. 409).[27]

We could go on like this, but it hardly seems helpful to get lost in debates and discussions that have been (and should be) lead elsewhere. In fact, Christopher von Bülow (2013), Douglas Roy (2017), Steve Stewart-Williams (2018), Dennett (2017, 2018), and Boudry (2018a, 2018b) have quite recently addressed several misunderstandings and popular objections against memes, including arguments against their causal effects, the aforementioned identification with neural substrates, issues with the notions of imitation and replication, and several others. Hence, we will now turn our attention to how memetics can be linked to other economically relevant evolutionary frameworks and, thereby, build bridges between those fields.

---

[26] In the literature, as an alternative to "retention," we can also find "reproduction" (e.g., Baraghith 2015), "transmission"' (e.g., Buskes 1998), "heredity" (e.g., Blackmore 2001, 2007; Dennett 1990), or "inheritance" (e.g., Hodgson and Knudsen 2010; Tang 2017), so that this algorithmic triad may be abbreviated to VSR, VST, VSH, or VSI, respectively.

[27] Remember that the German ethnologist Leo Frobenius—another forefather of memetics—already argued about a century ago that his basic tenet was to "understand culture as an organism independent from its human carriers [and] every cultural form as a living entity on its own ... First of all: cultures are not brought forth by human will, culture rather lives 'on' humans (Frobenius 1921, p. 3–4, own translation).

3.3 Implications of the Informationalist Perspective on Memes

3.3.1 No Meme is an Island: Memeplexes as Complex Systems

Several authors have used the notion of *co-adapted meme complexes* or, in short, *memeplexes* to describe a set of memes that replicate better together than on their own (Speel 1999; see also Blackmore 1999; Heylighen and Chielens 2009). Moreover, memes can be considered to coevolve with other memes as well as with genes and their phenotypic effects (Bull et al. 2000; Durham 1991, or Illies, 2005, 2010, on a related note). In short: "No meme is an island" (Dennett 1995, p. 144). This aspect can also be interpreted in the sense that memes may only gain meaning in connection with "context" (i.e., especially other memes). A similar argument has been advocated by John Hartley and Jason Potts, who regard the unit of cultural dynamics and evolution "… not as a discrete packet of replicating information, or indeed as any gene analogue, but rather as some locus of connection in a broader complex system …" (Hartley and Potts 2014, p. 119–120). Along these lines, Distin maintains that a "meme's own content may … be a fairly arbitrary factor in determining its success: its fortune in the struggle for survival *will always be relative to context*" (Distin 2005, p. 67, emphasis added). John Cook's so-called *meme complexes architecture* also goes in a similar direction, as he explains that memes can be both attracted to or repelled by clusters of memes (e.g., Cook 2008, 2015, with references to Fog 1999 and Graves 2005).

As already noted above, Tylor's definition mentions "that *complex* whole," memeplexes are an abbreviation of meme complexes, and Hartley and Potts explicitly refer to *complex systems*. It is, therefore, not very far fetched to view memes from a perspective that is also in line with *complexity science* (e.g., see Holland 2014; Miller 2015; or M. Mitchell 2009, for introductions). One important kind of complex systems was described by Hodgson and Knudsen (2006, 2010) as *complex population systems*:

> [C]omplex population systems contain multiple varied (intentional or non-intentional) entities that interact with the environment and each other. They face immediately scarce resources and struggle to survive, whether through conflict or cooperation. … They adapt and may pass on information to others, through replication or imitation (Hodgson 2011, p. 309).

With this in mind, we arrive at **proposition 2:** *Memes can be understood as entities of complex population systems. These entities interact with the environment and each other, face scarce resources, struggle to survive, and may pass on their information through replicative imitation by humans.*

Since one important systematic approach to complex systems is *network science* (Barabási 2016; Newman 2010), it is now plausible to depict memeplexes as *complex networks*, where nodes (or vertices) represent memes and edges (or links) depict a compatibility relation. More specifically, the connected memes are potentially the

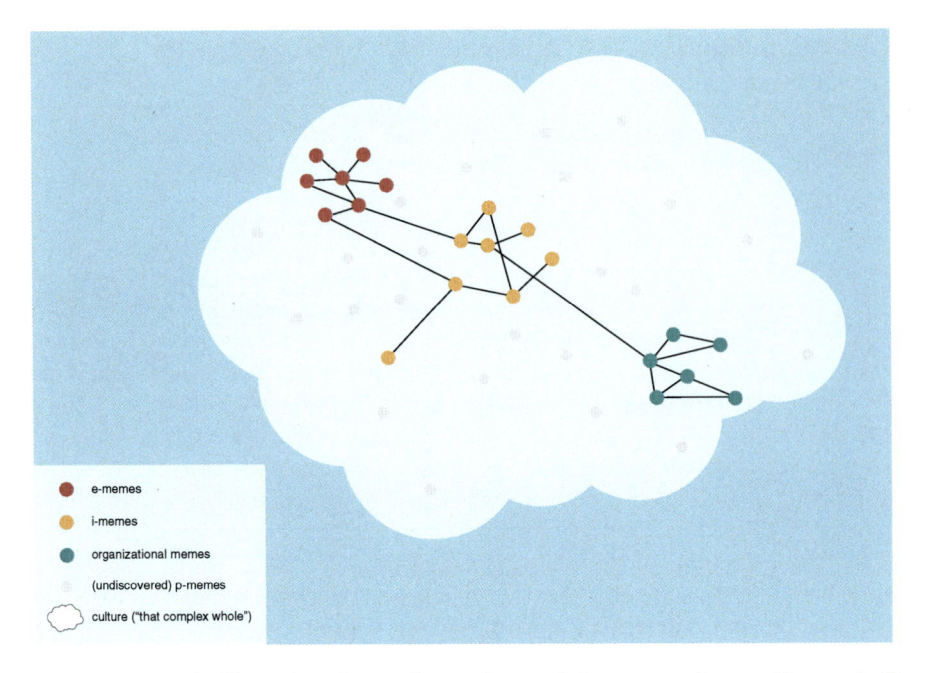

Fig. 3.2 (Schematic) illustration of a generic complex population system of memes ("memeplex") with different sub-populations. Memes are represented as nodes, and compatibility relations between memes are depicted by edges.

ones that positively reinforce the selection of those they are linked to. The edges may be undirected (for mutual compatibility) or directed (when compatibility or positive reinforcement is unidirectional). One advantage of this perspective is that it allows researchers to conceptualize (re-)combinations of memes from different cultural contexts such as the influence of political and religious meme(plexe)s (e.g., Protestantism) on economic ones (e.g., capitalism) as a graph.[28]

Figure 3.2 depicts this network approach to memeplexes as complex population systems in a schematic manner. The figure draws upon the p-i-e dimensions explained above as well as the notion of *organizational memes*, which have been defined by Sven Voelpel and his coauthors as "[a]ny of the core elements of organizational culture, like basic assumptions, norms, standards, and symbolic systems that can be transferred by imitation from one human mind to the next" (Voelpel et al 2005, p. 60). In this regard, Fig. 3.2 illustrates how various levels and sub-populations of these complex population systems may be linked by compatibility relations as proposed

---

[28]For Max Weber's famous but also controversial thesis on the influence of Protestantism on capitalism, see Weber (1930). Other examples may include the influence of basic paradigmatic (or "metaphysical") assumptions, ideas, concepts, and approaches from other scientific disciplines such as physics on economic ones and vice versa (e.g., Mirowski 1989; see also Schlaile et al. 2018a, with reference to Schramm 2016) or the memetic (in-)compatibility of sustainability-related worldviews and institutions (e.g., Harich 2010, 2015).

above. Note, however, that although organizational memes are represented as a distinct population here, they may also be distinguished into the mental representations (i-memes) of the organization's members and the environmental representations (e-memes) in terms of written documents, codes of conduct, business plans, practices, or verbalized expressions that in turn exhibit particular relations (e.g., Schlaile and Constantinescu 2016 and Chap. 4 below).

In summary, we may conclude this section with a suitable quote by Yulia Chentsova Dutton and Chip Heath:

> [T]he meme debate is just the latest example of a broader historical debate in the social sciences between proponents who see 'culture as elements' and those who see 'culture as a system' (Chentsova Dutton and Heath 2010, p. 50).

Arguably, a synthesis of those perspectives is called for and seems most promising also for economic applications, namely, to view memes as elements within complex (population) systems as proposed above.

3.3.2 The Economics of Attention: Memes in Competition for a Scarce Resource

Following proposition 2, an important question arises: Which "scarce resources" do memes face or compete for? Fortunately, Douglas Hofstadter already had an answer more than three decades ago: "Various mutations of a meme will have to compete with each other, as well as with other memes, for attention—which is to say, for brain resources in terms of both space and time devoted to that meme" (Hofstadter 1985, p. 51). In that regard, Distin adds another important point: "As memes struggle to gain and retain the attention of human minds, their success or failure is ... influenced more by the environment than by their own content" (Distin 2005, p. 67). In line with the above network approach, Distin further explains that "memetic selection will depend on memes' respective abilities to gain and retain our attention *in the current context:* fitness is always a relative concept" (ibid., p. 69, italics in original).

Before we can focus on this "resource of attention," another question immediately comes up: What is a meme's environment or context? The answer can be differentiated into (at least) two levels. First, there is the level of compatibility with other replicators (genes and memes) (level *a*). Second, there is the social network of the "carriers," i.e., human agents (level *b*). Successful transmission and selective retention of memes may, thus, depend on *a*) (in-)compatibility with these other replicators. Moreover, it can be assumed to depend on both b_1) the size of the (active) social network—which, in turn, is influenced by the information processing capabilities of the human brain (e.g., Dunbar 2011)—and b_2) the structure or topology and other properties of the social network (e.g., in terms of path length, degree distributions, clustering, strength of ties, centralities of agents, etc.) (see also Bogner 2019; Bogner et al. 2018). Moreover, there may also be feedback effects between levels *a* and *b* so that also (the structure of) the memeplex (*a*) has an effect upon the topology of the carrier network (b_2), for example, due to *cultural schemata* reinforcing homophily in the carrier network (Schlaile and Ehrenberger 2016).[29]

Coming back to attention, the "informationalist" perspective adopted above also brings us to Herbert Simon's famous argument that an information-rich world implies a scarcity of attention that necessitates the efficient allocation of attention among information sources (Simon 1971). The scientific discipline that deals with the allocation of scarce resources is economics; hence, it is also appropriate to speak of the *economics of attention* or an *attention economy* in this regard (see also Davenport and Beck 2001; Franck 1998; Falkinger 2007, 2008). According to Thomas Davenport and John Beck, attention can be defined as "focused mental engagement on a particular item of information" (Davenport and Beck 2001, p. 20, emphasis removed). With this definition in mind, I suggest **proposition 3**: *If memes are informational entities, they consume the attention of their carriers. Due to their integration into complex population systems of memes (memeplexes), the amount of attention they are allotted depends not only on the attractive power of their own informational content but also on their compatibility relations with other information sources (e.g., other memes in the system).*

3.4 Rules and Imitation

By taking another look at the definitions in Appendix: Selected Examples of Definitions or Uses of Memes, we can observe that some of them contain an explicit element of instruction (as already suggested by Cloak 1975) such as recipe or prescription (e.g., Dennett 2003, 2017), or rules and codes (e.g., Binmore 1998, Binmore and Samuelson 1994), which may be used as another link to an important strand of literature in evolutionary economics:[30] The so-called rule-based approach (RBA) to evolutionary economics has been developed mainly by Kurt Dopfer, John Foster, and Jason Potts (Dopfer 2001, 2004, 2005, 2016; Dopfer et al. 2004; Dopfer and Potts 2008, 2009, 2019; Potts 2019) and was recently applied to entrepreneurship by Georg Blind (2017). The *rule* is defined by Dopfer as "a *deductive schema that allows operations*" (Dopfer 2004, italics in original) and later by Dopfer and Potts as a "deductive procedure for operations" (Dopfer and Potts 2008, p. 104). According to Dopfer and Potts, a rule is "[w]hat the generic economic system is made of and what changes with economic evolution" (ibid.). Economic evolution is in turn understood as "a process that occurs at the *generic* level of the economic order, an analytic level that refers to the ideas, rules and knowledge, which constitute the basis of economic

---

[29] An illuminating review of (quite general) coevolutionary network dynamics is presented by Thilo Gross and Bernd Blasius (2008). A first operationalization of the propositions developed in this section can be found in Schlaile et al. (2018b) and Chap. 5 below.

[30] Interestingly, this aspect also reflects elements of the gene-like view because, as, e.g., Lee Kirkpatrick (2010, p. 72) notes, "[g]enes represent sets of instructions for building organisms (and their adaptations), not models of the final products."

operations" (Dopfer and Potts 2009, p. 24). According to the RBA, generic rules may be further distinguished into *subject* and *object* rules: subject rules are the cognitive and behavioral rules of an economic agent, while object rules are social and technical rules that represent the organizing principles for social and technological systems. The latter include, for instance, *Nelson–Winter organizational routines* and *Ostrom social rules* (e.g., Dopfer 2016; Dopfer and Potts 2019, for overviews). Arguably, cognitive rules constitute what I have called i-memes above. Without delving too much into rule taxonomies, it can already be argued that the RBA exhibits some striking similarities to and complementarities with memetic approaches. In fact, as Nobel laureate Elinor Ostrom, who also emphasized the importance of rules, already wrote:

> The central thesis that I have been pursuing is that rules are sets of instructions for creating an action situation … As such, rules are broadly analogous to genes, which are sets of instructions for creating a phenotype. Rules are memes rather than genes, but it is helpful to think about some of the similarities between genes and memes (Ostrom 2006, p. 116).

An important element of the RBA is the differentiation between micro, meso, and macro perspectives (see also Dopfer et al. 2016; Martin and Sunley 2007). The micro level is concerned with the individual agent or carrier dubbed *Homo Sapiens Oeconomicus*, which is conceived as a rule maker and rule user (Dopfer 2004). The meso level "focuses on single generic rules and their populations of actualizations" (Dopfer et al. 2004, p. 270), and the macro level concerns many rules and many populations of carriers. In the words of Dopfer and colleagues, "macro is not a behavioural aggregation of micro, but, rather, it offers a systems perspective on meso viewed as a whole" (Dopfer et al. 2004, p. 267).[31] This view is closely related to the notion of complex systems and approaches such as *agent-based modeling* (e.g., Wilenksy and Rand 2015), where local (behavioral) rules instead of a global "design plan" lead to the emergence of complex systemic behaviors such as a flock of birds (Dawkins 2009, Chap. 8).

According to the RBA, rules are made of ideas or information and also exhibit a process component that can be modeled as three phases of a meso trajectory comprised of *1. origination/emergence*, *2. diffusion (adoption and adaptation)*, and *3. retention* (Dopfer 2007; Dopfer and Potts 2008, 2009, 2019) (note a certain similarity to the variation-selection-retention triad mentioned above). As Georg Blind and Andreas Pyka lucidly summarize: "On the rule level of analysis [n.b.: which may be called the *rule's eye view*], a rule originates in one agent in the micro domain, diffuses into a population of agents that become its rule carriers in the meso domain, and eventually causes change in the rule structure of the macro domain" (Blind and Pyka 2014, p. 1087). However, as already Martin and Sunley (2007, p. 593) caution: "While this view [the RBA] has considerable potential, the precise meaning and content of such rules seems to require much further clarification and illustration."

Dopfer (2007, p. 35) explicitly mentions that rules (as ideas) always relate to other rules and have to fit in with them, which is equivalent to the argument in the

---

[31] In line with this differentiation, Ostrom also notes that "rules—as well as other cultural 'memes'—are likely to be selected at multiple levels" (Ostrom 2005, p. 30).

above Sect. 3.3.1 on compatibility relations and memeplexes. Remarkably enough, Francis Heylighen and Klaas Chielens (2009) also argue that a meme's structure can be analyzed as a *production rule*, which has the form "if condition, then action" (pp. 3209–3211). Especially in the case of these production rules, it is often a combination of rules which interact in a certain configuration of conditions ("if-rules") to generate the effect ("then-rule"), sometimes in combination with an alternative effect (an "else-rule"). However, it is important to emphasize that the differentiation, classification, and categorization of rules within the RBA is much more elaborate and detailed than in any memetic approach so far.

As already mentioned in Sect. 3.2, there exist at least two important imitation heuristics or rules-of-thumb, which can be regarded as overarching "meta-rules" particularly in conditions of uncertainty: imitate-the-successful and imitate-the-majority. In the words of Armen Alchian, "uncertainty provides an excellent reason for imitation of observed success" (Alchian 1950, p. 219).[32] Accordingly, we can argue that in the diffusion phase of the meso trajectory, one possible mechanism of adoption is the imitation of rules, thereby moving from micro (one rule, one carrier) to meso (a population of carriers with an actualization of the same rule). In this regard, a very simple example supports the intuition that rules (which can be called memes when being adopted via imitation) may often be "adaptive" or "maladaptive" not solely by content but by context: As long as you are the only driver on the road, it does not matter which side of the street you will be driving on. However, the rule for driving on the right side will obviously be "maladaptive" in a context where every other carrier (or at least the majority) drives on the left side. This is where our imitation heuristics come into play as an evolved approximation for evaluating the compatibility of a rule with the rules actualized in the population of the other carriers (in this example the other drivers). The essence of this simple example can be transferred to a broad variety of cultural and economic *institutions* (e.g., Patzelt 2007, 2012; Wäckerle 2014). As Hodgson (2007b, p. 8) writes: "Institutions are enduring systems of socially ingrained rules. They channel and constrain behaviour so that individuals form new habits as a result." Potts (2007, 2019) also champions an evolutionary institutional economics that draws upon the RBA and claims that "in evolutionary economics an institution is properly defined as an evolving process-structure of coordination as the evolutionary outcome of the process of a *generic rule* over a *meso trajectory*. An institution, then, is both a process and a structure" (Potts 2007, p. 343, italics in original).[33] I can now frame **proposition 4:** *Whenever rules become actualized in a population of carriers (meso) by imitation, these rules should be treated as an important subset or subtype of memes that can serve as "building blocks" for institutions.*

---

[32] However, imitate-the-majority is only an imperfect heuristic that is misleading and biased in the case where an agent's perceived majority does not coincide with the actual majority, a condition called "majority illusion" (Lerman et al. 2016).

[33] Brendan Markey-Towler (2019) develops this vision of institutional evolution further, unfortunately dismissing the relevance of memes altogether with one footnote.

Fig. 3.3 Institutions as the intersection of the sets of memes (M), habits (H), and rules (R)

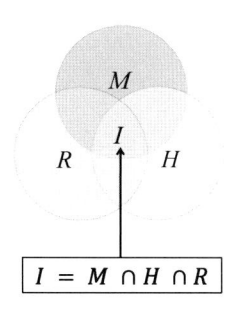

$$I = M \cap H \cap R$$

By taking the meme's eye view—and thus Dennett's (1995) famous "cui bono" question—seriously, however, we can clearly see why institutions are not necessarily evolving in the interest of society. As Alex Rosenberg puts it:

> The simple error functionalists made, which made their view sound so implausible, was to mis-identify the *beneficiaries* of the functions that institutions, practices, and organizations fulfilled. … They assumed, quite myopically and wrongly, that the function of institutions, practices, organizations, was to fulfill the needs of people, of human beings. But it was obvious that many institutions, practices, organizations are in fact harmful to people, confer no net advantage or benefit on them (Rosenberg 2017, p. 344, emphasis in original).

One could even go so far as to hypothesize that as long as rules do not become memes, they will play no role in economic evolution. However, these rule-memes, even if they are actualized in a large population of carriers, remain only *potential* institutional building blocks as long as they lack the component of stabilization that habits provide. In combination with proposition 1, we may thus picture economic institutions as the intersection of the sets of memes, habits, and rules as depicted in Fig. 3.3.

3.5 Conclusion: Building Bridges and Pointing the Way

While cultural evolutionists have sometimes, perhaps overzealously, claimed that a Darwinian approach may unify the social sciences in general (e.g., see Bagg 2017; Mesoudi 2011), the above discussion at least supports the idea that econometics may be useful for bridging some of the fragmented lines of research related to evolutionary economics. The prerequisite for econometics to do so is to acknowledge that not all definitions of the meme are equally suitable and that an "informationalist" view may be more adequate than a strict "neuralist" perspective as argued in Sect. 3.2. Moreover, although imitation is important, not every process of communication and social learning involves the exchange of information via replication, so that, at least in its current state, memetics is not—and probably should not even strive to become—an independent discipline. By establishing connections to various

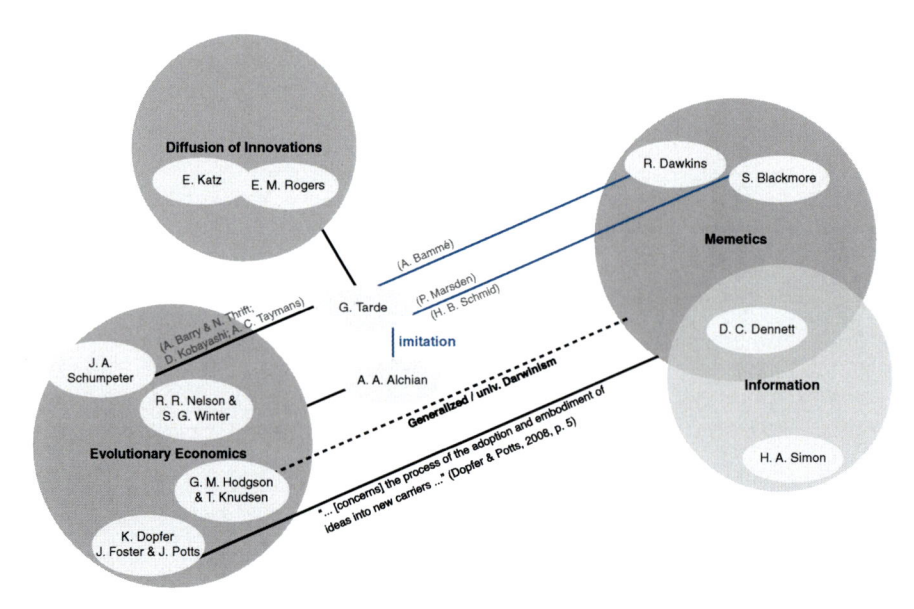

Fig. 3.4 Illustration of compatibility relations between selected approaches and authors in evolutionary economics, memetics, and related fields

related (or, compatible) fields, however, memetics can serve as a link between previously rather independent lines of research particularly by illuminating the ontological commonalities of the underlying research objects and approaches.

Figure 3.4 serves as a summary of some of the arguments in the previous sections. Here, I essentially take a "meta-memetic" perspective by illustrating how some of the disciplines could be linked in terms of their compatibility relations (as introduced in Sect. 3.3). Although Fig. 3.4 is just a subjective and nonexhaustive schematization, we can also observe that due to its connections to several fields, Tarde's work probably deserves much more attention by evolutionary economists than it currently gets.

To summarize, this conceptual paper contributes to the literature on evolutionary economics by exploring if taking up memetics again is actually as futile an effort as some authors have claimed it to be. For some critics, this endeavor may seem like "flogging a dead horse," but this discussion suggests otherwise: By taking some recent arguments from the memetics literature more seriously, we may actually contribute to closing two research gaps in evolutionary economics at the same time, namely, that imitation is underresearched and that cultural evolution is often neglected. The paper's starting point has been the general proposition that the (f)utility of integrating memetics with evolutionary economics should be re-evaluated. Based on an extensive review of critical arguments, some necessary clarifications, and categorizations, my first proposition establishes a connection to Hodgson and Knudsen's (e.g., 2010) habits. As a next step, I link to complex systems by means of my next two propositions that memes can be understood as elements of complex population systems that, as informational entities, particularly face the scarce resource of attention and need to be compatible with the existing memes in the system. These propositions result in the suggestion of a network approach to meme(plexe)s that is illustrated in Fig. 3.2, which calls for further development in the future. Another link is established by the discussion of the RBA and the rule component of several definitions of memes. In this way, I arrive at my fourth proposition that creates the link to evolutionary institutionalism (as suggested in Fig. 3.3 above).

On the whole, this contribution opens the door for new research endeavors along the lines of the network-scientific approach suggested in Sect. 3.3 and for further refinement or revision of my propositions. Indeed, an early attempt to operationalize the network approach can already be found elsewhere (Schlaile et al. 2018b; Chap. 5 below). Future research on the evolution of rules is also advisable, especially because the RBA and its micro–meso–macro differentiation offers much conceptual and explanatory potential also for the evolutionary literature beyond evolutionary economics. Finally, reviewing the literature has shown that (econo-)memetics still has to offer an emphasis on treating the meme rather than the carrier as the "beneficiary" of the evolutionary process, whereas even the recent evolutionary economics literature still appears to be biased towards the carriers and thereby disregard the "cui bono" point.

Appendix: Selected Examples of Definitions or Uses of Memes

The list is presented in alphabetical order and is by no means exhaustive. If not explicitly stated otherwise, the quote refers to the definition of a meme (or memes).

- Aunger (2002) claims that replicators must exist as specific (physical) substrates and defines the *neuromeme* as: "A configuration in one node of a neuronal network that is able to induce the replication of its state in other nodes" (p. 197, emphasis removed).
- Binmore (1998): "A meme is a norm, an idea, a rule of thumb, a code of conduct—something that can be replicated from one head to another by imitation or education, and which determines some aspects of the behavior of the person in whose head it is lodged" (p. 267).
- Binmore and Samuelson (1994): "Experience equips people with rules-of-thumb that they use to settle coordination problems that arise in real-life bargaining situations and elsewhere. ... We [refer] ... to such rules-of-thumb or codes-of-conduct as *memes*. We see a social norm as being one particular kind of meme" (p. 47, italics in original).
- Blackmore (1998): "If we define memes as transmitted by imitation then whatever is passed on by this copying process is a meme. Memes fulfil the role of replicator because they exhibit all three of the necessary conditions; that is, heredity (the form and details of the behaviour are copied), variation (they are copied with errors, embellishments or other variations), and selection (only some behaviours

are successfully copied)" (see also Blackmore 1999, p. 51, on an identical definition).
- Blackmore (2001): "We can define imitation as a process of copying that supports an evolutionary process, and define memes as the replicator which is transmitted when this copying occurs" (p. 229).
- Cloak (2015): "The meme is the unit of culture and of cultural transmission/acquisition. Memes are copied from brain to brain and, when activated, mediate behaviors which, in turn, result in social interactions, artifact creation, world views, and everything else we customarily refer to as 'cultural'" (Sect. 2.1).
- Csikszentmihalyi (1993): "Perhaps the best definition of a meme is 'any permanent pattern of matter or information produced by an act of human intentionality.' ... Memes come into being when the human nervous system reacts to an experience, and codes it in a form that can be communicated to others" (p. 120).
- Csikszentmihalyi (1998): "*Meme* has to do with imitation. ... Memes are units of information that we transmit by learning. In many ways memes obey the same pattern of variation, selection, transmission that genes do, but they are transmitted extrasomatically" (p. 210, italics in original). And "you can think of memes as being any kind of instruction for practice that you have learned from childhood on" (p. 211).
- Dawkins' (1976) original definition of the meme as "a unit of cultural transmission, or a unit of *imitation*" (Dawkins 2016, p. 249, italics in original).
- Dawkins (1982): "A meme should be regarded as a unit of information residing in a brain (Cloak's 'i-culture'). It has a definite structure, realized in whatever physical medium the brain uses for storing information" (p. 109, with reference to Cloak 1975).
- Delius (1991): "Synaptic patterns that code cultural traits will be called memes, by analogy with the molecular patterns that code biological traits and which are called genes" (p. 83). Hence, Delius pictures a "meme as a constellation of activated neuronal synapses lodged somewhere in the brain of an individual" (ibid.).
- Dennett (1995): "These new replicators are, roughly, ideas. Not the 'simple ideas' of Locke and Hume (the idea of red, or the idea of round or hot or cold), but the sort of complex ideas that form themselves into *distinct memorable units* ... [T]he units are the smallest elements that replicate themselves with reliability and fecundity" (p. 344, italics in original). More specifically, "what is preserved and transmitted in cultural evolution is *information*—in a media-neutral, language-neutral sense. Thus the meme is primarily a *semantic* classification, not a *syntactic* classification that might be directly observable in 'brain language' or natural language" (pp. 353–354, italics in original).
- Dennett (2003): "A meme is an information-packet with attitude—a recipe or instruction manual for doing something cultural. ... What is a meme made of? It is made of information, which can be carried in *any* physical medium" (p. 176, italics in original).
- Dennett (2017): "*Memes are informational things.* They are 'prescriptions' for ways of doing things that can be transmitted, stored, and mutated without being

executed or expressed (rather like recessive genes traveling silently in a genome)" (p. 211, italics in original).
- Distin (2014) argues that "memes might be units of representational content: cultural information preserved in a representational form that has a potential effect on or through those who acquire it" (p. 3). More specifically, "memes are … those bits of our mental 'furniture' that control our behaviour in response to the information that they carry, which we can link in our own minds to other such representations, and which preserve their content in a way that can be transmitted to other people" (pp. 3–4).
- Durham (1991) suggests "that 'meme' is a reasonable name for the functional unit of cultural transmission. … I take it to represent actual units of socially transmitted information, regardless of their form, size, and internal organization. The point is that whenever culture changes, *some* ideational unit is adopted and one or more homologous alternatives are not. That unit I will call a 'meme' …" (p. 189, italics in original).
- Esser (1999): "Memes are elements of knowledge and, hence, … cultural carriers of information and programs" (p. 203, own translation).
- Evers (1998): "Any meme can be defined generally as a *rule of behavior*, encoded by functional neuronal groups or pathways" (p. 439, italics in original).
- Finkelstein and Ayyub (2010): "A meme is information transmitted by any number of sources to at least an order of magnitude more recipients than sources, and propagated during at least twelve hours" (p. 303, emphasis removed). See also Finkelstein (2008).
- Flake (1998): "A unit of cultural information that represents a basic idea that can be transferred from one individual to another, and subjected to mutation, crossover, and adaptation" (p. 457, emphasis removed).
- Gatherer (1998): "An observable cultural phenomenon, such as a behaviour, artefact or an objective piece of information, which is copied, imitated or learned, and thus may replicate within a cultural system. Objective information includes instructions, norms, rules, institutions and social practices provided they are observable" (Sect. 9).
- Grant (1990): "A contagious information pattern that replicates by parasitically infecting human minds and altering their behavior, causing them to propagate the pattern. … An idea or information pattern is not a meme until it causes someone to replicate it …" (no pagination).
- Gunders and Brown (2010): "A meme is a cultural expression that is passed on from one person or group to another person or group" (p. 4, emphasis removed).
- Heylighen and Chielens (2009): "A cultural replicator; a unit of imitation or communication" (p. 3205). A replicator is in turn defined as an "information pattern that is able to make copies of itself, typically with the help of another system. Examples are genes, memes, and (computer) viruses" (ibid.).

- Kappelhoff (2012) uses memes to denote a short version of "behavior-guiding informational units of cultural evolution, e.g., in the form of knowledge, convictions, behavioral rules, and values … that are transmitted via imitation and social learning without implying a claim of analogy to genetic evolution" (p. 132, own translation).
- Moritz (1995): "A meme is an informational replicator whose principal attributes are pattern and meaning." (p. 158)
- Nye (2011): "A meme can be defined by its functional ability to sustainably reproduce within a society through social learning. This is similar to how computer viruses are defined as a subset of all possible combinations of code strings. In this view, a meme's semantic information contains a functional definition" (p. 15).
- Oxford Dictionaries (undated): "An element of a culture or system of behaviour passed from one individual to another by imitation or other non-genetic means."
- Patzelt (2000): "Memes are elements that become imprinting factors of culture-specific reality construction respectively institutionalization" (p. 76).
- Patzelt (2015a): "A meme is a cultural pattern that can be perceived and reproduced" (p. 171, own translation). Pattern is in turn used in the sense of Ruth Benedict (1934); see also Patzelt (2015b).
- Plotkin (1993) defines the meme as "the unit of cultural heredity analogous to the gene" (p. 251). "Memes are roughly equivalent to ideas or representations, that is, the internal end of the knowledge relationship" (p. 215).
- Price and Shaw (1998) define the meme as "the smallest element capable of being exchanged, with an associated sense of meaning and interpretation, to another brain" (p. 160, emphasis removed).
- Runciman (2009) defines memes as "items or packages of information transmitted from mind to mind by imitation or learning …" (p. 3).
- Schurz (2011, p. 210) associates himself with the view that memes are localized in brains: "Memes are … neuronal or mental structures, depending on whether one wants to reduce mental to neuronal or not" (own translation).
- Sheehan (2006) defines memes as "patterns that serve as templates for their own replication or translation" (pp. 9, 76, 260).
- Stanovich (2005) proposes "to view a meme as a brain control (or informational) state that can potentially cause fundamentally new behaviors and/or thoughts when replicated in another brain" (p. 175).
- Taylor and Giroux (2005) interpret "memes … as structures of the symbolic systems we employ as humans in constituting communication, and most centrally those of language" (p. 136).
- Wilkins (1998): "A meme is the least unit of sociocultural information relative to a selection process that has favourable or unfavourable selection bias that exceeds its endogenous tendency to change" (no pagination, emphasis removed).
- Wilson (1998): "We recommend that the unit of culture—now called meme—be the same as the node of semantic memory and its correlates in brain activity" (p. 148).

References

Acerbi, A., & Mesoudi, A. (2015). If we are all cultural Darwinians what's the fuss about? Clarifying recent disagreements in the field of cultural evolution. *Biology & Philosophy, 30*(4), 481–503.
Alchian, A. A. (1950). Uncertainty, evolution, and economic theory. *The Journal of Political Economy, 58*(3), 211–221.
Aldrich, H. E., Hodgson, G. M., Hull, D. L., Knudsen, T., Mokyr, J., & Vanberg, V. J. (2008). In defence of generalized Darwinism. *Journal of Evolutionary Economics, 18*, 577–596.
Almudi, I., & Fatas-Villafranca, F. (2018). Promotion and coevolutionary dynamics in contemporary capitalism. *Journal of Economic Issues, 52*(1), 80–102.
Aunger, R. (Ed.). (2000). *Darwinizing culture: The status of memetics as a science*. Oxford: Oxford University Press.
Aunger, R. (2002). *The electric meme: A new theory of how we think*. New York: Free Press.
Bagg, S. (2017). When will a Darwinian approach be useful for the study of society? *Politics, Philosophy & Economics, 16*(3), 259–281.
Baldwin, J. R., Faulkner, S. L., Hecht, M. L., & Lindsley, S. L. (Eds.). (2006). *Redefining culture: Perspectives across the disciplines*. Mahwah, NJ: Lawrence Erlbaum Associates.
Ball, J. A. (1984). Memes as replicators. *Ethology and Sociobiology, 5*, 145–161.
Bammé, A. (2009). Nicht Durkheim, sondern Tarde. Grundzüge einer anderen Soziologie. Nachwort. In A. Bammé (Ed.), *Gabriel Tarde - Die sozialen Gesetze* (pp. 109–153). Marburg: Metropolis.
Barabási, A.-L. (2016). *Network Science*. Cambridge: Cambridge University Press.
Baraghith, K. (2015). *Kulturelle Evolution und die Rolle von Memen: Ein Mehrebenenmodell*. Frankfurt a. M.: Peter Lang.
Barrett, L., Dunbar, R. I. M., & Lycett, J. (2002). *Human evolutionary psychology*. Basingstoke (Houndmills): Palgrave Macmillan.
Barry, A., & Thrift, N. (2007). Gabriel Tarde: Imitation, invention and economy. *Economy and Society, 36*(4), 509–525.
Beck, D. E., & Cowan, C. C. (1996). *Spiral dynamics: Mastering values, leadership, and change - Exploring the new science of memetics*. Malden, MA: Blackwell.
Becker, G. S. (1996). *Accounting for tastes*. Cambridge, MA: Harvard University Press.
Becker, M. C., & Knudsen, T. (2012). Nelson and Winter revisited. In M. Dietrich & J. Krafft (Eds.), *Handbook of the economics and theory of the firm* (pp. 243–255). Cheltenham: Edward Elgar.
Beinhocker, E. D. (2006). *The origin of wealth: Evolution, complexity, and the radical remaking of economics*. Boston, MA: Harvard Business School Press.
Beinhocker, E. D. (2011). Evolution as computation: Integrating self-organization with generalized Darwinism. *Journal of Institutional Economics, 7*(3), 393–423.
Benedict, R. (1934). *Patterns of culture*. London: Routledge.
Beugelsdijk, S., & Maseland, R. (2011). *Culture in economics: History, methodological reflections, and contemporary applications*. Cambridge: Cambridge University Press.
Binmore, K. (1998). *Game theory and the social contract II: Just playing*. Cambridge, MA: The MIT Press.
Binmore, K., & Samuelson, L. (1994). An economist's perspective on the evolution of norms. *Journal of Institutional and Theoretical Economics, 150*(1), 45–63.
Blackmore, S. (1998). Imitation and the definition of a meme. *Journal of Memetics—Evolutionary Models of Information Transmission, 2*. http://cfpm.org/jom-emit/1998/vol2/blackmore_s.html.
Blackmore, S. (1999). *The Meme Machine*. Oxford: Oxford University Press.
Blackmore, S. (2001). Evolution and memes: The human brain as a selective imitation device. *Cybernetics and Systems, 32*(1–2), 225–255.
Blackmore, S. (2003). The 'new science of memetics': The case for. *Think, 2*(3), 21–26.
Blackmore, S. (2005). Can memes meet the challenge? Susan Blackmore on Greenberg and on Chater. In S. Hurley & N. Chater (Eds.), *Perspectives on imitation: From neuroscience to social*

science, Imitation, human development, and culture (Vol. 2, pp. 409–411). Cambridge, MA: The MIT Press.

Blackmore, S. (2007). Memes, minds, and imagination. In I. Roth (Ed.), *Imaginative minds* (pp. 61–78). Proceedings of the British Academy. Oxford: Oxford University Press.

Blackmore, S. (2010). Memetics does provide a useful way of understanding cultural evolution. In F. J. Ayala & R. Arp (Eds.), *Contemporary debates in philosophy of biology* (pp. 225–272). Chichester: Wiley-Blackwell.

Blackmore, S. (2016). Memes and the evolution of religion: We need memetics, too. *Behavioral and Brain Sciences, 39*, 22–23.

Blind, G. D. (2017). *The entrepreneur in rule-based economics: Theory, empirical practice, and policy design.* Cham: Springer.

Blind, G., & Pyka, A. (2014). The rule-approach in evolutionary economics: A methodological template for empirical research. *Journal of Evolutionary Economics, 24*, 1085–1105.

Blute, M. (2005). Memetics and evolutionary social science. *Journal of Memetics—Evolutionary Models of Information Transmission, 9*. http://cfpm.org/jom-emit/2005/vol9/blute_m.html.

Blute, M. (2010). *Darwinian sociocultural evolution: Solutions to dilemmas in cultural and social theory.* Cambridge: Cambridge University Press.

Bogner, K., Mueller, M., & Schlaile, M. P. (2018). Knowledge diffusion in formal networks—The roles of degree distribution and cognitive distance. *International Journal of Computational Economics and Econometrics, 8*(3/4), 388–407.

Bogner, K. (2019). *United we stand, divided we fall: Essays on knowledge and its diffusion in innovation networks.* Doctoral dissertation, University of Hohenheim, Stuttgart. http://nbn-resolving.de/urn:nbn:de:bsz:100-opus-16151.

Boudry, M. (2018a). Invasion of the mind snatchers: On memes and cultural parasites. *Teorema, 37*(2), 111–124.

Boudry, M. (2018b). Replicate after reading: On the extraction and evocation of cultural information. *Biology & Philosophy, 33*.

Boudry, M., & Hofhuis, S. (2018). Parasites of the mind. Why cultural theorists need the meme's eye view. *Cognitive Systems Research, 52*, 155–167.

Boyd, R., & Richerson, P. J. (1985). *Culture and the evolutionary process.* Chicago: The University of Chicago Press.

Boyd, R., & Richerson, P. J. (2005). *The origin and evolution of cultures.* Oxford: Oxford University Press.

Boyer, P. (2018). *Minds make societies: How cognition explains the world humans create.* New Haven, CT: Yale University Press.

Breslin, D. (2010). Generalising Darwinism to study socio-cultural change. *International Journal of Sociology and Social Policy, 30*, 427–439.

Breslin, D. (2011). Reviewing a generalized Darwinist approach to studying socio-economic change. *International Journal of Management Reviews, 13*, 218–235.

Breslin, D. (2016). What evolves in organizational co-evolution? *Journal of Management & Governance, 20*(1), 45–67.

Brodie, R. (1996). *Virus of the mind. The new science of the meme.* Seattle: Integral Press.

Buenstorf, G. (2006). How useful is generalized Darwinism as a framework to study competition and industrial evolution? *Journal of Evolutionary Economics, 16*, 511–527.

Bull, L., Holland, O., & Blackmore, S. (2000). On meme-gene coevolution. *Artificial Life, 6*(3), 227–235.

Buskes, C. (1998). *The genealogy of knowledge: A Darwinian approach to epistemology and philosophy of science.* Tilburg: Tilburg University Press.

Buskes, C. (2013). Darwinism extended: A survey of how the idea of cultural evolution evolved. *Philosophia, 41*(3), 661–691.

Buskes, C. (2015). Darwinizing culture: Pitfalls and promises. *Acta Biotheoretica, 63*(2), 223–235.

Buskes, C. (2016). Light will be thrown: The emerging science of cultural evolution. *International Journal of Humanities, Art and Social Studies, 1*(1), 17–31.

Callebaut, W. (2011a). Beyond generalized Darwinism. I. Evolutionary economics from the perspective of naturalistic philosophy of biology. *Biological Theory, 6*(4), 338–350.
Callebaut, W. (2011b). Beyond generalized Darwinism. II. More things in heaven and earth. *Biological Theory, 6*(4), 351–365.
Campbell, D. T. (1960). Blind variation and selective retention in creative thought as in other knowledge processes. *Psychological Review, 67*(6), 380–400.
Campbell, D. T. (1965). Variation and selective retention in socio-cultural evolution. In H. R. Barringer, G. I. Blanksten, & R. W. Mack (Eds.), *Social change in developing areas: A reinterpretation of evolutionary theory* (pp. 19–49). Cambridge, MA: Schenkman.
Campbell, D. T. (1974). Evolutionary epistemology. In P. A. Schilpp (Ed.), *The philosophy of Karl Popper* (pp. 413–463). Lasalle, IL: Open Court.
Carroll, J., Clasen, M., Jonsson, E., Kratschmer, A. R., McKerracher, L., Riede, F., et al. (2017). Biocultural theory: The current state of knowledge. *Evolutionary Behavioral Sciences, 11*(1), 1–15.
Cartwright, J. (2008). *Evolution and human behavior: Darwinian perspectives on human nature* (2nd ed.). Cambridge, MA: The MIT Press.
Chentsova Dutton, Y., & Heath, C. (2010). Cultural evolution: Why are some cultural variants more successful than others? In M. Schaller, A. Norenzayan, S. J. Heine, T. Yamagishi, & T. Kameda (Eds.), *Evolution, culture, and the human mind* (pp. 49–70). New York: Taylor & Francis.
Chesterman, A. (2016). *Memes of translation: The spread of ideas in translation theory* (Revised ed.). Amsterdam: John Benjamins.
Claidière, N., Scott-Phillips, T. C., & Sperber, D. (2014). How Darwinian is cultural evolution? *Philosophical Transactions of the Royal Society B, 369.*
Cloak, F. T. (1975). Is a cultural ethology possible? *Human Ecology, 3*(3), 161–182.
Cloak, F. T. (2015). *A natural science of culture; or, a neurological model of the meme and of meme replication.* Version 3.2. https://www.tedcloak.com/a-natural-science-of-culture-32-beta.html.
Cook, J. E. (2008). *The role of the individual in organisational cultures: A Gravesian integrated approach.* Doctoral dissertation, Sheffield Hallam University.
Cook, J. E. (2015). Social and cultural influences on organisational change: The practical role of memeplexes. In T. Christensen (Ed.), *Innovative development* (pp. 230–255). Tucson, AZ: Integral Publishers.
Cordes, C. (2006). Darwinism in economics: From analogy to continuity. *Journal of Evolutionary Economics, 16*, 529–541.
Cordes, C. (2007a). Can a generalized Darwinism be criticized? A rejoinder to Geoffrey Hodgson. *Journal of Economic Issues, 41*(1), 277–281.
Cordes, C. (2007b). Turning economics into an evolutionary science: Veblen, the selection metaphor, and analogical thinking. *Journal of Economic Issues, 41*(1), 135–154.
Creanza, N., Kolodny, O., & Feldman, M. W. (2017). Cultural evolutionary theory: How culture evolves and why it matters. *PNAS, 114*(30), 7782–7789.
Cronk, L. (1999). *That complex whole: Culture and the evolution of human behavior.* Boulder, CO: Westview Press.
Csikszentmihalyi, M. (1993). *The evolving self: A psychology for the third millennium.* New York: HarperCollins.
Csikszentmihalyi, M. (1998). Self and evolution. *The NAMTA Journal, 23*(1), 205–233.
Cullen, B. (2000). *Contagious ideas: On evolution, culture, archaeology, and cultural virus theory. Collected writings* (J. Steele, R. Cullen, & C. Chippindale, Eds.). Oxford: Oxbow Books.
Davenport, T. H., & Beck, J. C. (2001). *The attention economy: Understanding the new currency of business.* Boston: Harvard Business School Press.
Dawkins, R. (1976). *The selfish gene* (1st ed.). Oxford: Oxford University Press.
Dawkins, R. (1982). *The extended phenotype: The long reach of the gene.* Oxford: Oxford University Press.
Dawkins, R. (1993). Viruses of the mind. In B. Dahlbohm (Ed.), *Dennett and his critics: Demystifying minds* (pp. 13–27). Oxford: Blackwell.

Dawkins, R. (2009). *The greatest show on earth: The evidence for evolution*. New York: Free Press.
Dawkins, R. (2015). *Brief candle in the dark: My life in science*. New York: HarperCollins.
Dawkins, R. (2016). *The selfish gene* (40th Anniversary ed.). Oxford: Oxford University Press.
Dawkins, R. (2017). Universal Darwinism. In G. Somerscales (Ed.), *Richard Dawkins: Science in the soul - Selected writings of a passionate rationalist* (pp. 119–150). London: Bantam Press.
Dawkins, R., & Wong, Y. (2016). *The ancestor's tale: A pilgrimage to the dawn of life* (2nd ed.). London: Orion Books.
Dawlabani, S. E. (2013). *Memenomics: The next-generation economic system*. New York: SelectBooks.
Delius, J. D. (1991). The nature of culture. In M. S. Dawkins, T. R. Halliday, & R. Dawkins (Eds.), *The Tinbergen legacy* (pp. 75–99). London: Chapman & Hall.
Dennett, D. C. (1990). Memes and the exploitation of imagination. *The Journal of Aesthetics and Art Criticism, 48*(2), 127–135.
Dennett, D. C. (1995). *Darwin's dangerous idea: Evolution and the meanings of life*. London: Simon & Schuster.
Dennett, D. C. (2002). The new replicators. In M. Pagel (Ed.), *Encyclopedia of evolution* (Vol. 1, pp. E83–E92). Oxford: Oxford University Press.
Dennett, D. C. (2003). *Freedom evolves*. New York: Penguin.
Dennett, D. C. (2006). *Breaking the spell: Religion as a natural phenomenon*. New York: Penguin.
Dennett, D. C. (2011). The evolution of culture. Originally published on edge.org, Feb. 1999: https://edge.org/conversation/the-evolution-of-culture. Reprinted in J. Brockman (Ed.), *Culture: Leading scientists explore societies, art, power, and technology* (pp. 1–26). New York: HarperCollins.
Dennett, D. C. (2013). *Intuition pumps and other tools for thinking*. New York: W. W. Norton.
Dennett, D. C. (2017). *From bacteria to Bach and back: The evolution of minds*. New York: W. W. Norton.
Dennett, D. C. (2018). Comment on Boudry. *Teorema, 37*(3), 125–127.
Distin, K. (2005). *The selfish meme. A critical reassessment*. Cambridge: Cambridge University Press.
Distin, K. (2011). *Cultural evolution*. Cambridge: Cambridge University Press.
Distin, K. (2014). *Foreword to the Chinese translation of The Selfish Meme*. https://distin.co.uk/kate/pdf/Foreword_Chinese.pdf.
Dollimore, D. E. (2006). *Darwinian evolutionary ideas in business economics and organization studies*. Doctoral dissertation, University of Hertfordshire. http://hdl.handle.net/2299/14978.
Dollimore, D. E. (2014). Untangling the conceptual issues raised in Reydon and Scholz's critique of organizational ecology and Darwinian populations. *Philosophy of the Social Sciences, 44*(3), 282–315.
Dopfer, K. (2001). Evolutionary economics - framework for analysis. In K. Dopfer (Ed.), *Evolutionary economics: Program and scope* (pp. 1–44). New York: Springer.
Dopfer, K. (2004). The economic agent as rule maker and rule user: Homo Sapiens Oeconomicus. *Journal of Evolutionary Economics, 14*, 177–195.
Dopfer, K. (2005). Evolutionary economics: A theoretical framework. In K. Dopfer (Ed.), *The evolutionary foundations of economics* (pp. 3–55). Cambridge: Cambridge University Press.
Dopfer, K. (2007). *Grundzüge der Evolutionsökonomie – Analytik, Ontologie und theoretische Schlüsselkonzepte*. Department of Economics Discussion Paper no. 2007–10. University of St. Gallen.
Dopfer, K. (2016). Evolutionary economics. In G. Faccarello & H. D. Kurz (Eds.), *Handbook on the history of economic analysis* (pp. 175–193). Cheltenham: Edward Elgar.
Dopfer, K., Foster, J., & Potts, J. (2004). Micro-meso-macro. *Journal of Evolutionary Economics, 14*, 263–279.
Dopfer, K., & Potts, J. (2008). *The general theory of economic evolution*. London: Routledge.
Dopfer, K., & Potts, J. (2009). On the theory of economic evolution. *Evolutionary and Institutional Economics Review, 6*(1), 23–44.

Dopfer, K., & Potts, J. (2019). Why is evolutionary economics not an empirical science? In F. Gagliardi & D. Gindis (Eds.), *Institutions and the evolution of capitalism. Essays in honour of Geoffrey M. Hodgson* (pp. 314–326). Cheltenham: Edward Elgar.

Dopfer, K., Potts, J., & Pyka, A. (2016). Upward and downward complementarity: The meso core of evolutionary growth theory. *Journal of Evolutionary Economics, 26,* 753–763.

Dugatkin, L. A. (2000). *The imitation factor: Evolution beyond the gene.* New York: Free Press.

Dunbar, R. I. M. (2011). Constraints on the evolution of social institutions and their implications for information flow. *Journal of Institutional Economics, 7*(3), 345–371.

Durham, W. H. (1991). *Coevolution: Genes, culture, and human diversity.* Stanford: Stanford University Press.

Esser, H. (1999). *Soziologie: Allgemeine Grundlagen* (3rd edn.). Frankfurt a. M.: Campus.

Essletzbichler, J., & Rigby, D. L. (2007). Exploring evolutionary economic geographies. *Journal of Economic Geography, 7*(5), 549–571.

Evers, J. R. (1998). A justification of societal altruism according to the memetic application of Hamilton's rule. In J. Ramaekers (Ed.), *Proceedings of the 15th international congress on cybernetics* (pp. 437–442). Namur: Association Internationale de Cybernétique.

Falkinger, J. (2007). Attention economies. *Journal of Economic Theory, 133,* 266–294.

Falkinger, J. (2008). Limited attention as a scarce resource in information-rich economies. *The Economic Journal, 118*(532), 1596–1620.

Finkelstein, R. (2008). *Defining memes.* http://www.semioticon.com/virtuals/memes2/finkelstein_paper.pdf.

Finkelstein, R., & Ayyub, B. M. (2010). Memetics for threat reduction in risk management. In J. G. Voeller (Ed.), *Wiley handbook of science and technology for homeland security* (pp. 301–309). Hoboken: Wiley.

Flake, G. W. (1998). *The computational beauty of nature: Computer explorations of fractals, chaos, complex systems, and adaptation.* Cambridge, MA: The MIT Press.

Fog, A. (1999). *Cultural selection.* Dordrecht: Kluwer Academic Publishers.

Fowers, B. J. (2015). *The evolution of ethics: Human sociality and the emergence of ethical mindedness.* Basingstoke: Palgrave Macmillan.

Franck, G. (1998). *Ökonomie der Aufmerksamkeit.* Munich: Hanser.

Frank, L. (2010). Evolutionäre Pädagogik und Memtheorie. In K. Gilgenmann, P. Mersch, & A. K. Treml (Eds.), *Kulturelle Vererbung* (pp. 141–174). Norderstedt: Books on Demand.

Frobenius, L. (1921). *Paideuma: Umrisse einer Kultur- und Seelenlehre.* Munich: C.H. Beck'sche Verlagsbuchhandlung.

Fuchs, S. (2001). Beyond agency. *Sociological Theory, 19*(1), 24–40.

Galbraith, J. K. (1998). *The affluent society.* 40th anniversary edition, updated and with a new introduction by the author. Boston: Mariner Books.

Gatherer, D. (1998). Why the 'Thought Contagion' metaphor is retarding the progress of memetics. *Journal of Memetics - Evolutionary Models of Information Transmission, 2.* http://cfpm.org/jom-emit/1998/vol2/gatherer_d.html

Geisendorf, S. (2009). The economic concept of evolution: Self-organization or universal Darwinism? *Journal of Economic Methodology, 16*(4), 377–391.

Gers, M. (2008). The case for memes. *Biological Theory, 3*(4), 305–315.

Gigerenzer, G., Hertwig, R., & Pachur, T. (Eds.). (2011). *Heuristics: The foundation of adaptive behavior.* Oxford: Oxford University Press.

Gladwell, M. (2000). *The tipping point: How little things can make a big difference.* Boston: Little, Brown and Company.

Godfrey-Smith, P. (2000). The replicator in retrospect. *Biology and Philosophy, 15,* 403–423.

Godfrey-Smith, P. (2009). *Darwinian populations and natural selection.* Oxford: Oxford University Press.

Godfrey-Smith, P. (2014). *Philosophy of biology.* Princeton: Princeton University Press.

Gorodnichenko, Y., & Roland, G. (2011). Which dimensions of culture matter for long-run growth? *American Economic Review, 101*(3), 492–498.

Grant, G. (1990). *Memetic lexicon*. http://pespmc1.vub.ac.be/MEMLEX.html.
Graves, C. W. (2005). *The never ending quest: A treatise on an emergent cyclical conception of adult behavioral systems and their development* (C. C. Cowan & N. Todorovic, Eds.). Santa Barbara, CA: ECLET.
Gross, T., & Blasius, B. (2008). Adaptive coevolutionary networks: A review. *Journal of the Royal Society Interface, 5*(20), 259–271.
Guiso, L., Sapienza, P., & Zingales, L. (2006). Does culture affect economic outcomes? *Journal of Economic Perspectives, 20*(2), 23–48.
Gunders, J., & Brown, D. (2010). *The complete idiot's guide to memes: Find out what makes an idea catch on*. New York: Penguin.
Hannon, E., & Lewens, T. (2014). Cultural evolution. In J. B. Losos (Ed.), *The Princeton guide to evolution* (pp. 795–800). Princeton: Princeton University Press.
Harich, J. (2010). Change resistance as the crux of the environmental sustainability problem. *System Dynamics Review, 26*(1), 35–72.
Harich, J. (2015). Solving difficult large-scale social system problems with root cause analysis. *Spanda Journal, 6*(1), 53–66.
Hartley, J., & Potts, J. (2014). *Cultural science: A natural history of stories, demes, knowledge and innovation*. London: Bloomsbury.
Henrich, J., & Gil-White, F. J. (2001). The evolution of prestige: Freely conferred deference as a mechanism for enhancing the benefits of cultural transmission. *Evolution and Human Behavior, 22*(3), 165–196.
Herrmann-Pillath, C. (2000). *Evolution von Wirtschaft und Kultur: Bausteine einer transdisziplinären Methode*. Marburg: Metropolis.
Herrmann-Pillath, C. (2010a). *The economics of identity and creativity: A cultural science approach*. New Brunswick: Transaction.
Herrmann-Pillath, C. (2010b). What have we learnt from 20 years of economic research into culture? *International Journal of Cultural Studies, 13*(4), 317–335.
Herrmann-Pillath, C. (2013). *Foundations of economic evolution: A treatise on the natural philosophy of economics*. Cheltenham: Edward Elgar.
Heylighen, F., & Chielens, K. (2009). Evolution of culture, memetics. In R. A. Meyers (Ed.), *Encyclopedia of complexity and systems science* (pp. 3205–3220). New York: Springer.
Hodgson, G. M. (2000). What is the essence of institutional economics? *Journal of Economic Issues, 34*(2), 317–329.
Hodgson, G. M. (2002). Darwinism in economics: From analogy to ontology. *Journal of Evolutionary Economics, 12*(3), 259–281.
Hodgson, G. M. (2005). Generalizing Darwinism to social evolution: Some early attempts. *Journal of Economic Issues, 39*(4), 899–914.
Hodgson, G. M. (2007a). A response to Christian Cordes and Clifford Poirot. *Journal of Economic Issues, 41*(1), 265–276.
Hodgson, G. M. (2007b). Introduction. In G. M. Hodgson (Ed.), *The evolution of economic institutions: A critical reader* (pp. 1–15). Cheltenham: Edward Elgar.
Hodgson, G. M. (2009). The nature and replication of routines. In M. C. Becker & N. Lazaric (Eds.), *Organizational routines: Advancing empirical research* (pp. 26–44). Cheltenham: Edward Elgar.
Hodgson, G. M. (2011). A philosophical perspective on contemporary evolutionary economics. In J. B. Davis & D. W. Hands (Eds.), *The Elgar companion to recent economic methodology* (pp. 299–318). Cheltenham: Edward Elgar.
Hodgson, G. M. (2013a). Clarifying generalized Darwinism: A reply to Scholz and Reydon. *Organization Studies, 34*(7), 1001–1005.
Hodgson, G. M. (2013b). Understanding organizational evolution: Toward a research agenda using generalized Darwinism. *Organization Studies, 34*(7), 973–992.
Hodgson, G. M., & Knudsen, T. (2006). Why we need a generalized Darwinism, and why generalized Darwinism is not enough. *Journal of Economic Behavior & Organization, 61*(1), 1–19.

Hodgson, G. M., & Knudsen, T. (2008). In search of general evolutionary principles: Why Darwinism is too important to be left to the biologists. *Journal of Bioeconomics, 10*, 51–69.

Hodgson, G. M., & Knudsen, T. (2010). *Darwin's conjecture: The search for general principles of social and economic evolution*. Chicago: University of Chicago Press.

Hodgson, G. M., & Knudsen, T. (2011). Generalizing Darwinism and evolutionary economics: From ontology to theory. *Biological Theory, 6*, 326–337.

Hodgson, G. M., & Knudsen, T. (2012). Agreeing on generalised Darwinism: A response to Pavel Pelikan. *Journal of Evolutionary Economics, 22*(1), 9–18.

Hodgson, G. M., & Lamberg, J.-A. (2018). The past and future of evolutionary economics: Some reflections based on new bibliometric evidence. *Evolutionary and Institutional Economics Review, 15*(1), 167–187.

Hofstadter, D. R. (1985). *Metamagical themas: Questing for the essence of mind and pattern*. New York: Basic Books.

Holland, J. H. (2014). *Complexity: A very short introduction*. Oxford: Oxford University Press.

Hull, D. L. (1988a). *Science as a process: An evolutionarry account of the social and conceptual development of science*. Cambridge, MA: The MIT Press.

Hull, D. L. (1982). The naked meme. In H. C. Plotkin (Ed.), *Learning, development, and culture: Essays in evolutionary epistemology* (pp. 273–327). Chichester: Wiley.

Hull, D. L. (1988b). Interactors versus vehicles. In H. C. Plotkin (Ed.), *The role of behavior in evolution* (pp. 19–50). Cambridge, MA: MIT Press.

Illies, C. (2005). Die Gene, die Meme und wir. In B. Goebel, A. M. Hauk, & G. Kruip (Eds.), *Probleme des Naturalismus* (pp. 127–160). Paderborn: Mentis.

Illies, C. (2010). Biologie statt Philosophie? Evolutionäre Kulturerklärungen und ihre Grenzen. In J. Oehler (Ed.), *Der Mensch - Evolution, Natur und Kultur* (pp. 213–231). Berlin: Springer.

Ingold, T. (2013). Prospect. In T. Ingold & G. Palsson (Eds.), *Biosocial becomings: Integrating social and biological anthropology* (pp. 1–21). Cambridge: Cambridge University Press.

Iwai, K. (1984). Schumpeterian dynamics: An evolutionary model of innovation and imitation. *Journal of Economic Behavior and Organization, 5*(2), 159–190.

Jagers op Akkerhuis, G. A. J. M., Spijkerboer, H. P., & Koelewijn, H. -P. (2016). Generalising Darwinian evolution by using its smallest-scale representation as a foundation. In G. A. J. M. Jagers op Akkerhuis (Ed.), *Evolution and transitions in complexity: The science of hierarchical organization in nature* (pp. 103–123). Cham: Springer.

Jesiek, B. K. (2003). *Betwixt the popular and academic: The histories and origins of memetics*. Master's thesis, Virginia Polytechnic Institute and State University, Blacksburg. http://hdl.handle.net/10919/42774

Jones, E. L. (1995). Culture and its relationship to economic change. *Journal of Institutional and Theoretical Economics, 151*(2), 269–285.

Kappelhoff, P. (2012). Selektionsmodi der Organisationsgesellschaft: Gruppenselektion und Memselektion. In S. Duschek, M. Gaitanides, W. Matiaske, & G. Ortmann (Eds.), *Organisationen regeln: Die Wirkmacht korporativer Akteure* (pp. 131–162). Wiesbaden: Springer.

Katz, E. (2006). Rediscovering Gabriel Tarde. *Political Communication, 23*(3), 263–270.

Keen, S. (2011). *Debunking economics - revised and expanded edition. The naked emperor dethroned?*. London: Zed Books.

Kinnunen, J. (1996). Gabriel Tarde as a founding father of innovation diffusion research. *Acta Sociologica, 39*(4), 431–442.

Kirkpatrick, L. A. (2010). From genes to memes: Psychology at the nexus. In M. Schaller, A. Norenzayan, S. J. Heine, T. Yamagishi, & T. Kameda (Eds.), *Evolution, culture, and the human mind* (pp. 71–79). New York: Psychology Press.

Klasing, M. J. (2013). Cultural dimensions, collective values and their importance for institutions. *Journal of Comparative Economics, 41*(2), 447–467.

Knudsen, T. (2008). Organizational routines in evolutionary theory. In M. C. Becker (Ed.), *Handbook of organizational routines* (pp. 125–151). Cheltenham: Edward Elgar.

Kobayashi, D. (2014). Effects of anthropology and archaeology upon early innovation studies.

Kobayashi, D. (2015a). *Invention and development: Toward Schumpeter's early innovation theory.* http://hdl.handle.net/2115/60236.

Kobayashi, D. (2015b). Schumpeter as a diffusionist: A new interpretation of Schumpeter's theory of socio-cultural evolution. *Evolutionary and Institutional Economics Review, 12*(2), 265–281.

Lakoff, G., & Johnson, M. (1980). *Metaphors we live by.* Chicago: The University of Chicago Press.

Latour, B. (2002). Gabriel Tarde and the end of the social. In P. Joyce (Ed.), *The social in question* (paperback ed. 2014, pp. 117–132). Abingdon: Routledge.

Lawson, T. (2006). The nature of heterodox economics. *Cambridge Journal of Economics, 30,* 483–505.

Lee, F. (2009). *A history of heterodox economics: Challenging the mainstream in the twentieth century.* London: Routledge.

Lerman, K. (2016). Information is not a virus, and other consequences of human cognitive limits. *Future Internet, 8*(2), 21.

Lerman, K., Yan, X., & Wu, X.-Z. (2016). The "majority illusion" in social networks. *PLoS ONE, 11*(2), e0147617.

Lewens, T. (2007). *Darwin.* London: Routledge.

Lewens, T. (2015). *Cultural evolution: Conceptual challenges.* Oxford: Oxford University Press.

Lewens, T. (2019). Cultural evolution. In E. N. Zalta (Ed.), *The Stanford encyclopedia of philosophy* (Summer 2019). https://plato.stanford.edu/archives/sum2019/entries/evolution-cultural/

Lynch, A. (1996). *Thought contagion: How belief spreads through society. The new science of memes.* New York: Basic Books.

Mahner, M., & Bunge, M. (1997). *Foundations of biophilosophy.* Berlin: Springer.

Markey-Towler, B. (2019). The competition and evolution of ideas in the public sphere: A new foundation for institutional theory. *Journal of Institutional Economics, 15*(1), 27–48.

Marsden, P. (2000). Forefathers of memetics: Gabriel Tarde and the laws of imitation. *Journal of Memetics - Evolutionary Models of Information Transmission, 4.* http://cfpm.org/jom-emit/2000/vol4/marsden_p.html.

Martin, R., & Sunley, P. (2007). Complexity thinking and evolutionary economic geography. *Journal of Economic Geography, 7*(5), 573–601.

McCraw, T. K. (2007). *Prophet of innovation. Joseph Schumpeter and creative destruction.* Cambridge, MA: The Belknap Press of Harvard University Press.

McNamara, A. (2011). Can we measure memes? *Frontiers in Evolutionary Neuroscience, 3.*

Mesoudi, A. (2007). A Darwinian theory of cultural evolution can promote an evolutionary synthesis for the social sciences. *Biological Theory, 2*(3), 263–275.

Mesoudi, A. (2011). *Cultural evolution. How Darwinian theory can explain human culture and synthesize the social sciences.* Chicago: University of Chicago Press.

Mesoudi, A. (2016a). Cultural evolution: A review of theory, findings and controversies. *Evolutionary Biology, 43*(4), 481–497.

Mesoudi, A. (2016b). Cultural evolution: Integrating psychology, evolution and culture. *Current Opinion in Biotechnology, 7,* 17–22.

Mesoudi, A. (2017). Pursuing Darwin's curious parallel: Prospects for a science of cultural evolution. *PNAS, 114*(30), 7853–7860.

Mesoudi, A., Whiten, A., & Laland, K. N. (2006). Towards a unified science of cultural evolution. *Behavioral and Brain Sciences, 29*(4), 329–383.

Miller, J. H. (2015). *A crude look at the whole: The science of complex systems in business, life and society.* New York: Basic Books.

Mirowski, P. (1989). *More heat than light: Economics as social physics: Physics as nature's economics.* Cambridge: Cambridge University Press.

Mitchell, M. (2009). *Complexity: A guided tour.* Oxford: Oxford University Press.

Mitchell, P. (2012). *Contagious metaphor.* London: Bloomsbury.

Morin, O. (2016). *How traditions live and die.* Oxford: Oxford University Press.

Moritz, E. (1995). MetaSystem Transitions, memes, and cybernetic immortality. *World Futures, 45,* 155–171.

Müller, S. S. W. (2010). *Theorien sozialer Evolution: Zur Plausibilität darwinistischer Erklärungen sozialen Wandels*. Bielefeld: transcript.
Murdock, G. P. (1960). How culture changes. In H. L. Shapiro (Ed.), *Man, culture, and society* (pp. 247–260). New York: Oxford University Press.
Murdock, G. P. (1965). *Culture and society*. Pittsburgh: University of Pittsburgh Press.
Nau, H. (2004). Reziprozität, Eliminierung oder Fixierung? Kulturkonzepte in den Wirtschaftswissenschaften im Wandel. In G. Blümle, N. Goldschmidt, R. Klump, B. Schauenberg, & H. von Senger (Eds.), *Perspektiven einer kulturellen Ökonomik* (pp. 249–269). Münster: LIT.
Nelson, R. R. (2006). Evolutionary social science and universal Darwinism. *Journal of Evolutionary Economics, 16*, 491–510.
Nelson, R. R. (2007). Universal Darwinism and evolutionary social science. *Biology and Philosophy, 22*, 73–94.
Nelson, R. R., Dosi, G., Helfat, C., Pyka, A., Saviotti, P. P., Lee, K., et al. (2018). *Modern evolutionary economics: An overview*. Cambridge: Cambridge University Press.
Nelson, R. R. (2001). Evolutionary perspectives on economic growth. In K. Dopfer (Ed.), *Evolutionary economics: Program and scope* (pp. 165–194). New York: Springer.
Nelson, R. R., & Winter, S. G. (1982). *An evolutionary theory of economic change*. Cambridge, MA: The Belknap Press of Harvard University Press.
Newman, M. E. J. (2010). *Networks: An introduction*. Oxford, etc.: Oxford University Press.
Niosi, J. (2012). *Innovation and development through imitation (In praise of imitation)*. Paper presented at the meeting of the International Schumpeter Society, July 2-5. Brisbane.
Niosi, J. (2017). Imitation and innovation new biologics, biosimilars and biobetters. *Technology Analysis & Strategic Management, 29*(3), 251–262.
Nye, B. D. (2011). Modeling memes: A memetic view of affordance learning. *Publicly accessible Penn Dissertations. 336*. http://repository.upenn.edu/edissertations/336/
Ostrom, E. (2005). *Understanding institutional diversity*. Princeton: Princeton University Press.
Ostrom, E. (2006). The complexity of rules and how they may evolve over time. In C. Schubert & G. von Wangenheim (Eds.), *Evolution and design of institutions* (pp. 100–122). London: Routledge.
Oxford Dictionaries. (undated). Definition of meme in English. http://www.oxforddictionaries.com/definition/english/meme
Patzelt, W. J. (2000). Institutions as knowledge-gaining systems: What can social scientists learn from evolutionary epistemology? *Evolution and Cognition, 6*(1), 70–83.
Patzelt, W. J. (Ed.). (2007). *Evolutorischer Institutionalismus: Theorie und exemplarische Studien zu Evolution, Institutionalität und Geschichtlichkeit*. Würzburg: Ergon.
Patzelt, W. J. (Ed.). (2012). *Parlamente und ihre Evolution: Forschungskontext und Fallstudien*. Baden-Baden: Nomos.
Patzelt, W. J. (2015a). Der Schichtenbau der Wirklichkeit im Licht der Memetik. In B. P. Lange & S. Schwarz (Eds.), *Die menschliche Psyche zwischen Natur und Kultur* (pp. 170–181). Lengerich: Pabst.
Patzelt, W. J. (2015b). Was ist "Memetik"? In B. P. Lange & S. Schwarz (Eds.), *Die menschliche Psyche zwischen Natur und Kultur* (pp. 52–61). Lengerich: Pabst.
Pelikan, P. (2011). Evolutionary developmental economics: How to generalize Darwinism fruitfully to help comprehend economic change. *Journal of Evolutionary Economics, 21*, 341–366.
Persky, J. (1995). Retrospectives: The ethology of homo economicus. *The Journal of Economic Perspectives, 9*(2), 221–231.
Plotkin, H. C. (1993). *Darwin machines and the nature of knowledge* (paperback, 1997). Cambridge, MA: Harvard University Press.
Polichak, J. W. (1998). Memes - what are they good for? A critique of memetic approaches to information processing. *Skeptic, 6*(3), 45–53.
Popper, K. R. (1972). *Objective knowledge: An evolutionary approach*. Oxford: Oxford University Press.
Popper, K. R. (1974). Replies to my critics. In P. A. Schilpp (Ed.), *The philosophy of Karl Popper* (pp. 961–1197). Lasalle, IL: Open Court.

Potts, J. (2007). Evolutionary institutional economics. *Journal of Economic Issues, 41*(2), 341–350.

Potts, J. (2019). *Innovation commons: The origin of economic growth.* Oxford: Oxford University Press.

Price, I., & Shaw, R. (1998). *Shifting the patterns: Breaching the memetic codes of corporate performance.* Chalford: Management Books 2000.

Reydon, T. A. C. (2016). A critical assessment of graph-based generalized Darwinism. In G. A. J. M. Jagers op Akkerhuis (Ed.), *Evolution and transitions in complexity: The science of hierarchical organization in nature* (pp. 125–135). Cham: Springer.

Reydon, T. A. C., & Scholz, M. (2015). Searching for Darwinism in generalized Darwinism. *The British Journal for the Philosophy of Science, 66*(3), 561–589.

Richerson, P. J., & Christiansen, M. H. (Eds.). (2013). *Cultural evolution: Society, technology, language, and religion.* Cambridge, MA: The MIT Press.

Robert, V., Yoguel, G., & Lerena, O. (2017). The ontology of complexity and the neo-Schumpeterian evolutionary theory of economic change. *Journal of Evolutionary Economics, 27,* 761–793.

Rogers, E. M. (2003). *Diffusion of innovations* (5th ed.). New York: Simon and Schuster.

Rosenberg, A. (2017). Why social science is biological science. *Journal for General Philosophy of Science, 48*(3), 341–369.

Roy, D. (2017). Myths about memes. *Journal of Bioeconomics, 19*(3), 281–305.

Runciman, W. G. (2002). Heritable variation and competitive selection as the mechanism of sociocultural evolution. In M. Wheeler, J. Ziman, & M. A. Boden (Eds.), *The evolution of cultural entities. Proceedings of the British Academy* (Vol. 112, pp. 9–25). Oxford: Oxford University Press.

Runciman, W. G. (2009). *The theory of cultural and social selection.* Cambridge: Cambridge University Press.

Runciman, W. G. (2015). Evolutionary sociology. In J. H. Turner, R. Machalek, & A. Maryanski (Eds.), *Handbook on evolution and society: Toward an evolutionary social science* (pp. 194–214). Abingdon: Routledge.

Schlaile, M. P. (2013). *A 'more evolutionary' approach to economics: The Homo sapiens oeconomicus and the utility maximizing meme.* Paper presented at the 11th Globelics International Conference on entrepreneurship, innovation policy and development in an era of increased globalisation, September 11–13. Middle East Technical University, Ankara, Turkey.

Schlaile, M. P., & Constantinescu, L. (2016). Exploring the potential of organizational memetics: A review and case example. *Academy of Management Proceedings.* https://doi.org/10.5465/AMBPP.2016.17407abstract.

Schlaile, M. P., & Ehrenberger, M. (2016). Complexity, cultural evolution, and the discovery and creation of (social) entrepreneurial opportunities: Exploring a memetic approach. In E. S. C. Berger & A. Kuckertz (Eds.), *Complexity in entrepreneurship, innovation and technology research: Applications of emergent and neglected methods* (pp. 63–92). Cham: Springer.

Schlaile, M. P., Mueller, M., Schramm, M., & Pyka, A. (2018a). Evolutionary economics, responsible innovation and demand: Making a case for the role of consumers. *Philosophy of Management, 17*(1), 7–39.

Schlaile, M. P., Zeman, J., & Mueller, M. (2018b). It's a match! Simulating compatibility-based learning in a network of networks. *Journal of Evolutionary Economics, 28*(5), 1111–1150.

Schmid, H. B. (2004). Evolution by imitation. Gabriel Tarde and the limits of memetics. *Distinktion: Scandinavian Journal of Social Theory, 5*(2), 103–118.

Scholz, M., & Reydon, T. A. C. (2013). On the explanatory power of generalized Darwinism: Missing items on the research agenda. *Organization Studies, 34*(7), 993–999.

Schramm, M. (2008). *Ökonomische Moralkulturen. Die Ethik differenter Interessen und der plurale Kapitalismus.* Marburg: Metropolis.

Schramm, M. (2016). Wie funktioniert die Geschäftswelt wirklich? Business Metaphysics und Theorie der Firma. *ETHICA Wissenschaft und Verantwortung, 24*(4), 311–360.

Schubert, C. (2014). 'Generalized Darwinism' and the quest for an evolutionary theory of policymaking. *Journal of Evolutionary Economics, 24,* 479–513.

Schumpeter, J. A. (2002). The economy as a whole: Seventh chapter of The Theory of Economic Development, translated by Ursula Backhaus. *Industry and Innovation, 9*(1/2), 93–145. Originally published 1912.

Schurz, G. (2011). *Evolution in Natur und Kultur: Eine Einführung in die verallgemeinerte Evolutionstheorie*. Heidelberg: Spektrum.

Sheehan, E. L. (2006). *The mocking memes: A basis for automated intelligence*. Bloomington: AuthorHouse.

Shenkar, O. (2010a). *Copycats: How smart companies use imitation to gain strategic edge*. Boston: Harvard Business School Press.

Shenkar, O. (2010b). Defend your research: Imitation is more valuable than innovation. *Harvard Business Review*, April, 28–29.

Shennan, S. (2002). *Genes, memes and human history: Darwinian archaeology and cultural evolution*. London: Thames & Hudson.

Shiller, R. J. (2017). Narrative economics. *American Economic Review, 107*(4), 967–1004.

Shionoya, Y. (2008). Schumpeter and evolution: An ontological exploration. In Y. Shionoya & T. Nishizawa (Eds.), *Marshall and Schumpeter on evolution: Economic sociology of capitalist development* (pp. 15–35). Cheltenham: Edward Elgar.

Simon, H. A. (1971). Designing organizations for an information-rich world. In M. Greenberger (Ed.), *Computers, communication, and the public interest* (pp. 37–72). Baltimore, MD: Johns Hopkins Press.

Smith, C. M., Gabora, L., & Gardner-O'Kearny, W. (2018). The extended evolutionary synthesis facilitates evolutionary models of culture change. *Cliodynamics, 9*(2), 84–107.

Sparrow, T., & Hutchinson, A. (Eds.). (2013). *A history of habit: From Aristotle to Bourdieu*. Lanham, MD: Lexington.

Speel, H. -C. (1998). Memes are also interactors. In J. Ramaekers (Ed.), *Proceedings of the 15th international congress on cybernetics* (pp. 402–407). Association Internationale de Cybernétique.

Speel, H.-C. (1999). Memetics: On a conceptual framework for cultural evolution. In F. Heylighen, J. Bollen, & A. Riegler (Eds.), *The evolution of complexity: The violet book of "Einstein meets Magritte"* (pp. 229–254). Dordrecht: Kluwer Academic Publishers.

Spolaore, E., & Wacziarg, R. (2013). How deep are the roots of economic development? *Journal of Economic Literature, 51*(2), 325–369.

Stanovich, K. E. (2005). *The robot's rebellion: Finding meaning in the age of Darwin*. Chicago: The University of Chicago Press.

Stewart-Williams, S. (2018). *The ape that understood the universe: How the mind and culture evolve*. Cambridge: Cambridge University Press.

Stoelhorst, J. W. (2008a). Darwinian foundations for evolutionary economics. *Journal of Economic Issues, 42*(2), 415–423.

Stoelhorst, J. W. (2008b). The explanatory logic and ontological commitments of generalized Darwinism. *Journal of Economic Methodology, 15*(4), 343–363.

Stoelhorst, J. W. (2014). The future of evolutionary economics is in a vision from the past. *Journal of Institutional Economics, 10*(4), 665–682.

Tang, S. (2017). Toward generalized evolutionism: Beyond "generalized Darwinism" and its critics. *Journal of Economic Issues, 51*(3), 588–612.

Tarde, G. (1890). *Les lois de l'imitation: étude sociologique*. Paris: Félix Alcan.

Tarde, G. (1903). *The laws of imitation*. transl. by E. C. Parsons. New York: Henry Holt.

Taylor, J. R., & Giroux, H. (2005). The role of language in self-organizing. In G. A. Barnett & R. Houston (Eds.), *Advances in self-organizing systems* (pp. 131–167). Cresskill, NJ: Hampton Press.

Taymans, A. C. (1950). Tarde and Schumpeter: A similar vision. *The Quarterly Journal of Economics, 64*(4), 611–622.

Thomas, R. (2018). The claims of generalized Darwinism. *Philosophy of Management, 17*(2), 149–167.

Throsby, C. D. (2001). *Economics and culture*. Cambridge: Cambridge University Press.

Tylor, E. B. (1871). *Primitive culture. Researches into the development of mythology, philosophy, religion, language, art, and custom* (6th ed.). London: John Murray.
Veblen, T. (1899). *The theory of the leisure class: An economic study of institutions.* New York: The Macmillan Company.
Voelpel, S. C., Leibold, M., & Streb, C. K. (2005). The innovation meme: Managing innovation replicators for organizational fitness. *Journal of Change Management, 5*(1), 57–69.
von Bülow, C. (2013). Meme. [English translation of the (German) article "Mem". In J. Mittelstraß (Ed.), *Enzyklopädie Philosophie und Wissenschaftstheorie* (2nd edn., Vol. 5, pp. 318–324). Stuttgart: Metzler]. https://www.philosophie.uni-konstanz.de/typo3temp/secure_downloads/87495/0/de0f56268a8ad66b13cfc7652e092ce47ea79fb6/meme.pdf
von Sydow, M. (2012). *From Darwinian metaphysics towards understanding the evolution of evolutionary mechanisms: A historical and philosophical analysis of gene-Darwinism and universal Darwinism.* Göttingen: Universitätsverlag Göttingen.
Vromen, J. J. (1995). *Economic evolution: An enquiry into the foundations of new institutional economics.* London: Routledge.
Wäckerle, M. (2014). *The foundations of evolutionary institutional economics: Generic institutionalism.* London: Routledge.
Waddington, C. H. (1977). *Tools for thought.* St Albans: Paladin.
Weber, M. (1930). *The Protestant ethic and the spirit of capitalism.* Transl. by T. Parsons; with an introduction by A. Giddens. London: Routledge.
Wegener, F. (2015). *Memetik. Der Krieg des neuen Replikators gegen den Menschen* (3rd ed.). Gladbeck: Kulturförderverein Ruhrgebiet e.V.
Weng, L. (2014). *Information diffusion on online social networks.* Doctoral dissertation, School of Informatics and Computing, Indiana University. http://lilianweng.github.io/papers/weng-thesis-single.pdf.
Weng, L., Menczer, F., & Ahn, Y. -Y. (2013). Virality prediction and community structure in social networks. *Scientific Reports, 3.*
Weng, L., Menczer, F., & Ahn, Y. -Y. (2014). Predicting successful memes using network and community structure. In *Proceedings of the eighth international AAAI conference on weblogs and social media* (pp. 535–544). AAAI Press, Cambridge, MA.
Whiten, A., Hinde, R. A., Stringer, C. B., & Laland, K. N. (Eds.). (2012). *Culture evolves.* Oxford: Oxford University Press.
Wilensky, U., & Rand, W. (2015). *An introduction to agent-based modeling: Modeling natural, social, and engineered complex systems with NetLogo.* Cambridge, MA: MIT Press.
Wilkins, D. J. (1998). What's in a meme? Reflections from the perspective of the history and philosophy of evolutionary biology. *Journal of Memetics - Evolutionary Models of Information Transmission, 2.* http://cfpm.org/jom-emit/1998/vol2/wilkins_js.html
Wilkins, D. J., & Hull, D. L. (2014). Replication and reproduction. In E. N. Zalta (Ed.), *The Stanford encyclopedia of philosophy* (Spring 2014). https://plato.stanford.edu/archives/spr2014/entries/replication/
Wilson, E. O. (1998). *Consilience: The unity of knowledge.* New York: Random House.
Winter, S. G. (2017). Pursuing the evolutionary agenda in economics and management research. *Cambridge Journal of Economics, 41*(3), 721–747.
Witt, U. (1999). Bioeconomics as economics from a Darwinian perspective. *Journal of Bioeconomics, 1*(1), 19–34.
Witt, U. (2014). The future of evolutionary economics: Why the modalities of explanation matter. *Journal of Institutional Economics, 10*(4), 645–664.
Witt, U. (2016). *Rethinking economic evolution: Essays on economic change and its theory.* Cheltenham: Edward Elgar.
Wortmann, H. (2010). *Zum Desiderat einer Evolutionstheorie des Sozialen: Darwinistische Konzepte in den Sozialwissenschaften.* Constance: UVK Verlagsgesellschaft mbH.
Ziman, J. (Ed.). (2000). *Technological innovation as an evolutionary process.* Cambridge: Cambridge University Press.

Chapter 4
It's More Than Complicated! Using Organizational Memetics to Capture the Complexity of Organizational Culture

Michael P. Schlaile, Kristina Bogner, and Laura Mülder

Abstract While organizational and business researchers have fruitfully applied evolutionary theory at various levels of analysis, few utilize organizational memetics to capture the complexity of organizational culture. This article contributes to bridging the gap between theorizing and empirical research on organizational memetics by raising and addressing the question if and how the diversity and interdependence of organizational memes can be captured. To tackle this exploratory question, the authors present a comprehensive literature review on organizational memetics and demonstrate how meme mapping can be used to highlight interdependencies among organizational memes based on the case of a German consulting firm. Besides revealing the most prominent memes in the complex memetic system of the organization, the meme map illustrates connections of varying strength among the organizational

---

This chapter has been previously published and should be cited as Schlaile, M. P., Bogner, K., & Muelder, L. (2019). It's more than complicated! Using organizational memetics to capture the complexity of organizational culture. *Journal of Business Research* (article in press). doi: 10.1016/j.jbusres.2019.09.035. The authors are grateful to the employees of P3 automotive GmbH for their participation in the interviews. Moreover, the authors would like to thank Raul J. Kraus for establishing necessary contacts and for helpful suggestions. Special thanks are also due to Elisabeth Berger for a constructive pre-submission review, and to Gianpaolo Abatecola, Marion Büttgen, Matteo Cristofaro, Charles Galunic, Bijoy Goswami, Danny Gutknecht, Ilfryn Price, Vincenzo Uli, and J. T. Velikovsky for helpful discussions, comments, suggestions, and criticism. Earlier drafts of this paper were presented in 2016 at the annual conference of the European Academy of Management in Paris and the Annual Meeting of the Academy of Management in Anaheim, CA (DOI: 10.5465/ambpp.2016.17407abstract). Thus, the authors are also grateful to the reviewers and participants of both conferences for contributing to the evolution of this article. Last but not least, the authors thank the two anonymous referees for the Journal of Business Research for their constructive comments. All remaining errors and omissions are the authors' responsibility.

M. P. Schlaile (✉) · K. Bogner
Department of Innovation Economics, University of Hohenheim, Stuttgart, Germany
e-mail: schlaile@uni-hohenheim.de

K. Bogner
e-mail: kristina.bogner@uni-hohenheim.de

L. Mülder
Alumna, University of Hohenheim, Stuttgart, Germany
e-mail: c.laura88@googlemail.com

© The Author(s), under exclusive license to Springer Nature Switzerland AG 2021
M. P. Schlaile (ed.), *Memetics and Evolutionary Economics*, Economic Complexity and Evolution, https://doi.org/10.1007/978-3-030-59955-3_4

memes, thereby supporting the argument that organizational memetics can help to expose attractive memes that are important for both the stability and change of organizational cultures.

4.1 Introduction

Despite growing awareness among researchers that the complexity of business processes requires abandoning mechanistic reductionism in favor of acknowledging dynamic interactions and interrelated elements (e.g., Allen et al. 2011; Arthur 2015; Kirman 2011; McKelvey 1997; Wilson and Kirman 2016; Woodside 2017), there appears to be no consensus in the scientific discourse on the meaning of complexity (e.g., Boulton et al. 2015; Burnes 2005; Edmonds 1999; Jacobs 2013). Even in the somewhat narrower context of organizational complexity, Price (2004, p. 40) observes that "complexity has become an umbrella under which advocates of whole system approaches to issues of organizational change and development can gather." Consequently, in this line of research, business organizations have frequently been conceived as complex systems (e.g., Bandte 2007; Hazy et al. 2007; Lissack 1999; Malik 2016). The complex systems view has, in turn, a close relationship with evolutionary theory and approaches (e.g., Bar-Yam 2004; Kauffman 1993, 1995; Price 1999a, 2004; Wilson and Kirman 2016), including computational methods such as agent-based modeling (Bandte 2007; Beinhocker 2006; Breslin 2014; Schlaile et al. 2018b; Wilensky and Rand 2015).

In this article, a particular (Darwinian) evolutionary perspective is taken up in order to capture the complexity of organizational culture, namely, meme theory or memetics (e.g., Aunger 2000; Blackmore 1999; Dawkins 1976; Dennett 2017). For the purpose of this introductory section, the *Oxford Dictionary of English* definition of the term meme is adopted, which is "an element of a culture or system of behaviour passed from one individual to another by imitation or other non-genetic means" (https://en.oxforddictionaries.com/definition/meme; see also Stevenson 2010). Correspondingly, memetics is simply "the study of memes" (according to the *Merriam-Webster* dictionary, https://www.merriam-webster.com/dictionary/memetics). Memetics has already received considerable attention from marketing researchers (e.g., Atadil et al. 2017; Hamlin et al. 2015; Marsden 1998, 2002; Murray et al. 2014; Vos and Varey 2012; Williams 2002, 2004; Wu and Ardley 2007). However, in view of the fact that memes are often considered to be constituent elements of culture, much unexploited conceptual and empirical potential exists regarding the combination of memetics with research on organizational culture (Russ 2014) and change processes in organizations and other complex social systems (Cook 2015; Waddock 2015, 2016, 2019). Whereas organizational and business researchers have fruitfully applied evolutionary theory at various levels of analysis (e.g., Abatecola et al. 2016; Aldrich and Ruef 2006; Baum and Singh 1994; Breslin 2016; Dosi and Marengo 2007; Galunic and Weeks 2005; Kumbartzki 2002; Murmann et al. 2003; Powell and Wakeley 2003; for an overview, see also Abatecola

2014), few make use of organizational memetics. Organizational memetics is a subfield of evolutionary organizational studies that adopts a memetic perspective (e.g., Gill 2012, 2013; Lord 2012; Manikandan 2009; Price 1995, 2009; Price and Shaw 1998; Shepherd 2002; Weeks and Galunic 2003). Of course, many alternative terms exist for memes as units of (organizational) culture or the mental programs of an organization, a non-exhaustive selection of which can already be found in Hofstede (1998, p. 478). However, despite the existence of other candidates for organizational replicators (e.g., Breslin 2016, for an overview), the present article follows the intellectual history of organizational memetics and uses the term meme in order to stress that the study is in line with a dynamic and process-oriented non-reductionistic framework (see also Price 1999a, for a related argument). After all, as Price (1999b) explains, a "combination of memetics and complexity … offers rich potential as we examine not individual memes passed on through imitation, but rich and densely interconnected memetic patterns" (Price 1999b, Sect. 2, with reference to Gabora 1997 and Price 1995, 1999a).

In view of the broad literature base, the article's contribution is twofold: (1) to present the current state of the literature on organizational memetics, and (2) to advance the interplay between theorizing and empirical research in this field. More precisely, the authors address the following exploratory research question: *(How) can the diversity and interdependence of characteristic elements of an organizational culture be captured such that these elements can be considered as representations of the underlying organizational memes?* At this stage, the relatively broad notion of organizational memes suggested by Voelpel et al. (2005, p. 60) is adopted as a working definition: "Any of the core elements of organizational culture, like basic assumptions, norms, standards, and symbolic systems that can be transferred by imitation from one human mind to the next." This article makes a first step toward tackling the empirical part of this question by presenting a small exploratory case study of a German engineering consulting firm with a remarkable organizational culture (e.g., remarkable with an eye to complexity due to the firm's informal structure with flat hierarchies and a self-conception as a dynamic, self-organizing network of diverse, interconnected actors).

The article is structured as follows: The subsequent Sect. 4.2 provides a brief recapitulation of the argument for viewing organizational culture from a perspective combining complexity theory and meme theory. Section 4.3 presents the literature review, which focuses on key contributions in the field of organizational memetics. This literature review serves as a starting point for Sect. 4.4, which introduces the case example. More precisely, Sect. 4.4 presents the results of a small qualitative study based on data gathered via semi-structured in-depth interviews. Building on these results, the authors highlight important aspects of the company's organizational culture that can be interpreted as relevant organizational memes, which are visualized by means of a meme mapping technique known from marketing literature (e.g., Atadil et al. 2017). The final Sect. 4.5 contains the conclusion, which summarizes this article's contribution and potential directions for further research.

4.2 Memetics and Complexity

Memes do not exist in isolation. Although many proponents of meme theory have focused on the particular characteristics of memes that may increase their success in replication (e.g., Heylighen and Chielens 2009), the notion of so-called *memeplexes* (e.g., Blackmore 1999; Speel 1999) takes account of the fact that the selection of memes strongly depends on compatibility with already existing memes in the system (e.g., Heylighen and Chielens 2009; Schlaile et al. 2018b; Weeks and Galunic 2003). Consistent with an understanding of organizational culture as a complex system (e.g., Frank and Fahrbach 1999), it is appropriate to follow Price (1999a) and Price and Shaw (1998) by regarding organizational culture as a complex pattern of memes or a *complex memetic system* (CMS). For the purpose of this article, a CMS shall be considered as a special type of *complex adaptive system*. As Price puts it:

> A CMS perspective would interpret socially constructed phenomena as, themselves, being emergent, self-organizing, memetic effects. Culture, dominant patterns of language and thinking, may be interpreted as one of the highest levels of memetic pattern (Price 1999a, p. 172).

According to Johnson, a complex system "contains a collection of many interacting objects or 'agents'" (Johnson 2007, p. 13). These agents "are linked together through their interactions, they can also be thought of as forming part of a network" (Johnson 2007, p. 13). This is why "for many scientists in the community, the study of Complexity is synonymous with the study of agents and networks together" (Johnson 2007, p. 13). In a CMS, the "agents" or objects are memes, and the "interactions" or links are the compatibility relations of a memeplex (e.g., Schlaile et al. 2018b). Regularly, a complex system exhibits path dependence and feedback that affect the objects' behavior such that the past affects the present, and events at one location of the system affect what happens at another part (e.g., Johnson 2007; Holland 2014). Moreover, a complex system "is typically open" (Johnson 2007, p. 14), that is, the system can and will be influenced by its environment, leading to a situation in which the "system evolves in a highly non-trivial and often complicated way, driven by an ecology of agents who interact and adapt under the influence of feedback" (Johnson 2007, pp. 14–15). In this regard, complex systems exhibit (occasionally extreme and surprising) *emergent phenomena*—as described by the common phrase "the whole is more than the sum of the parts" (e.g., Holland 2014, p. 4)—which are not directly controlled, or not even controllable, by a central planner or manager. Note how this understanding of a *complex* system goes beyond the colloquial use of the term complex as a synonym to *complicated* (hence the title of this article; see also Price 2004, on a related note).

Arguably, many properties of complex (adaptive) systems known from the literature (e.g., Holland 2014; Jacobs 2013; Sack 2015) apply to organizational culture, such as *diversity* or heterogeneity of components (i.e., organizational culture is made of diverse cultural elements), *non-linear interconnections* (i.e., the elements of an organizational culture interact or are connected in non-additive ways), *aggregation* and *hierarchy* (e.g., in the sense that simple building blocks form sub-systems, which are themselves building blocks for larger parts of the organizational culture),[1] *adaptation* (i.e., the components of an organizational culture may change in response to the state or behavior of other cultural elements), and *emergence* (i.e., the overall culture has features that cannot be derived from the features of individual cultural elements). While all of these properties are interesting and important in and of themselves, the study presented in Sect. 4.4 mainly focuses on the diversity and the interconnections of different memes in the CMS of an organizational culture.

4.3 Literature Review

An extensive review of the *general* literature on memes is beyond the scope of this article, keeping in mind that contributions and criticisms are widely dispersed among myriads of books and scientific journals from various different disciplines, including the now terminated *Journal of Memetics—Evolutionary Models of Information Transmission* (1997–2005).[2] Nevertheless, some introductory remarks are appropriate. Reviews (and critics) of memetics often focus on Dawkins' (1976) introduction of the meme as a new replicator in the first edition of *The Selfish Gene*. However, as Dawkins (2015) himself notes, significant advances in memetics have been made by others, including Aunger (2002), Blackmore (1999), and Dennett (1995, 2017). For an overview of memetics and a summary of some of the major criticisms, the reader is referred to Aunger (2000), Heylighen and Chielens (2009), and Dennett (2017). One of these criticisms is that memetic theory has been insufficiently subjected to empirical tests (e.g., Chielens and Heylighen 2005), a point that is being addressed in the meantime first and foremost by researchers studying the spread and evolution of memes on the Internet (e.g., Adamic et al. 2016; Schlaile et al. 2018a).

Various scholars have also started to explore the explanatory potential with respect to memetics in an organizational context, thereby giving rise to the still relatively unknown field of *organizational memetics,* which is the focus of this article.[3] The origin of the term organizational memetics can be attributed to Price (1995). In his article, he compares organizational learning with organic evolution by natural selection, positing that "organizational evolution (learning) can be considered as a selection process between mental replicators" (Price 1995, p. 299). In this conceptual article, Price draws on a wide range of literature from complexity theory to evolutionary biology and defines memes as "a composite mindset, a paradigm, or a mental model" (Price 1995, p. 307). Price's (1995) article is one of the first

---

[1] As Williams (2004, p. 776) puts it: "Memes … are subject to a hierarchy of memes, a social construction of beliefs. Interpersonal fitness is therefore determined … also by institutional fit."

[2] See http://cfpm.org/jom-emit/ for an archive of all articles published in that journal.

[3] Due to the simple fact that organizational memetics is not used consistently as a keyword by all relevant publications and because many publications using a memetic approach do not focus on organizational issues, the following review relies on a manual selection by the authors based on a combination of prior knowledge of the literature, additional keyword searches, and the classical "snowball technique" (Ridley 2012).

attempts to utilize memetics in an organizational context. Although the definitions and (empirical) implications remain rather vague, Price's (1995) article provides a seminal argument for using a combination of complexity theory and memetics to shed light on organizational evolution.

In 2001, the first empirical approach to organizational memetics was published, namely the article by Lord and Price (2001). Lord and Price draw upon the similarity between abiotic and biotic complex systems and employ a "phenetic analysis of memetic data" to achieve a "successful phylogenic reconstruction of the known pattern of descent of the main post-reformation Christian Churches" (ibid.). Their article constitutes a first proof of feasibility for empirical organizational memetics and already points to the fact that some organizational memes may be more important or more powerful than others.

Two years later, Weeks and Galunic (2003) published their groundbreaking article, which can be regarded as a major theoretical advancement in the field of organizational memetics: Weeks and Galunic propose a memetic theory of the firm that "provides a new perspective … on the question of why we have the firms that we have" (Weeks and Galunic 2003, p. 1309). More precisely, Weeks and Galunic eschew an instrumental or strictly functionalist interpretation of organizational culture by taking up the meme's eye view, thus stressing that firms do not exist because they are necessarily good for their members or for society but "fundamentally because they are good ways for memes to replicate themselves" (Weeks and Galunic 2003, p. 1321). The term meme is used by Weeks and Galunic (2003, p. 1344) as an umbrella term to "refer collectively to cultural modes of thought (ideas, beliefs, assumptions, values, interpretative schema, and know-how)." Their article sheds light on various shortcomings of alternative (especially functionalist) theories of the firm and on how a memetic perspective can be useful for organizational studies by focusing on memes as the relevant units in cultural evolution that compete for the scarce resource of attention. Consistent with complexity theory and the notion of CMS, organizational culture is regarded by Weeks and Galunic (2003) as an emergent phenomenon, the evolution of which strongly depends on existing combinations of memes and cannot be consciously designed in a top-down manner by means of fiat.

The article by Voelpel et al. (2005) also adopts the view of organizations as complex adaptive systems and focuses on companies' abilities to "embed and leverage an innovation culture in their organizations" (Voelpel et al. 2005, p. 57). They propose the *innovation meme* to be a central construct that can be used to identify and leverage the replicators of an organizational innovation culture. Moreover, they suggest definitions and give examples for an *organizational meme* and an *innovation meme* (reproduced in Table 4.2 in the appendix). Recognizing organizational complexity, Voelpel et al. (2005) put forward approaches to memetic engineering that take account of the self-organizing and emergent character of the organizational cultural system, which "does not provide much space for planned intervention through traditional managerial approaches" (Voelpel et al. 2005, p. 65). However, several parts of their approach remain rather vague, and the authors themselves acknowledge that their article should be seen as the "basis for further refinement, development, and testing of the concept …, and as the starting point for further scholarly debate and investigations" (Voelpel et al. 2005, p. 68).

The next fundamental advancement of organizational memetics can be seen in the empirical contribution by Shepherd and McKelvey (2009). Their study is consistent with a CMS perspective by regarding organizations as hosts of populations of memes that self-organize and affect organizational adaptation. The article fathoms "whether understanding organizational memetic variation is empirically possible" (Shepherd and McKelvey 2009, p. 135) by using an exploratory quasi-experimental design. For the purpose of their article, memes are defined as "independent knowledge-based units of meaning that can be (socially) exchanged—*transmitted*—with more or less accurate transfer with or without alteration of meaning" (Shepherd and McKelvey 2009, p. 138, italics in original). Their article can be regarded as an important step in organizational memetics since "the potential of a theory based on organizational memes in coevolution with the environment has been postulated, but remains empirically under-developed" (Shepherd and McKelvey 2009, p. 135), an assessment that still holds true today (see also Price 2012a, on a related note).

Table 4.2 in the appendix gives a chronological overview of contributions to the field of organizational memetics, including those outlined above as well as additional ones that have been omitted above for the sake of brevity. Table 4.2 reproduces various definitions and applications of the term meme throughout the respective contributions. Thereby, one can easily see that, thus far, no agreement has been reached on how memes are to be defined in an organizational context. In Gill's (2012) words, one can get the impression that "on each occasion memes are couched in terms which suit the message of the thesis rather than a consensus of what might constitute a putatively real entity" (Gill 2012, p. 326). However, another possibility is that the definitions shown in Table 4.2 do indeed represent different aspects or angles of memes so that this terminological ambiguity does not render the whole endeavor futile—as some critics may suggest. Nevertheless, the vast majority of articles is of a theoretical or conceptual nature with very few of the contributions represented in Table 4.2 containing actual empirical studies. Therefore, this article also presents an empirical case example in the following Sect. 4.4. By the same token, this empirical part can be read as another attempt to heed Hull's advice that memeticists "should shift away from general discussions toward attempts to apply these terms to real cases" (Hull 2000, p. 48). However, in anticipation of the results, it should be remembered that the following case example mainly serves the purpose of demonstrating how organizational memetics may be useful for capturing the complexity of an organizational culture.

4.4 Case Example

4.4.1 Sample and Method

For the purpose of this exploratory study, memes are operationalized as meaningful and compatible *units of knowledge* (see also Schlaile et al. 2018b) that can be regarded as constituent elements, that is, building blocks, of an organizational culture. In Balkin's (1998, p. 43) words: "Memes are the *building blocks* of the cultural software that forms our apparatus of understanding" (emphasis added). Essentially, in the present approach, an organizational culture is understood as a CMS of organizational memes that govern social interactions (see also Stoelhorst and Richerson 2013, p. S50, on a related note). Conceptually, this study draws upon the trichotomy or three-dimensional view on memes ("p-i-e") proposed by Schlaile and Ehrenberger (2016, pp. 68–69), and the authors acknowledge that this study does not measure the mental representation (i-meme) of a meme but an aspect of organizational culture by capturing an artifact, behavioral entity, or environmental representation of a meme (e-meme) in the form of a verbal expression. Note that, as Hofstede (1998, p. 479) already explained two decades ago: "Culture is a characteristic of the organization, not of individuals, but it is manifested in and measured from the verbal and/or nonverbal behaviour of individuals" (emphasis removed).

Qualitative data were collected between January and June 2014 via semi-structured in-depth interviews (in German) at the German consultancy firm *P3 automotive GmbH* (henceforth P3 automotive). Interviewees were selected mainly on the basis of seniority which resulted in a purposive sample with an average duration of employment of 9.18 years.[4] The sample includes various functional levels and ranks such as consultants, executives, and partners. Since culture is arguably one of the most central topics in the company, the authors observed a high willingness to cooperate with 80% of the initially contacted potential interviewees actually agreeing to the request. Other than their interest in the topic, interviewees had no incentive for participation. The final sample includes a total of 16 interviews with an average duration of about 60 min, ten of which were conducted face-to-face, six via telephone. The interviewees were anonymized by means of assignable four-digit codes ranging from IP01 to IP16.

The data gathered through the interviews were analyzed in seven steps, the first six of which take account of the special requirements associated with *qualitative content analysis* (e.g., Kuckartz 2014, 2019; Mayring 2000, 2014), whereas the seventh step combines meme mapping methods (Atadil et al. 2017) with social network and brand analysis (Fronzetti Colladon 2018):

1. Transcribing conversations based on recordings and notes,
2. creating memory minutes,
3. inductive text categorization and segmentation and searching for relevant information,
4. relating relevant information to categories (and translating from German to English),
5. determining frequencies of mentioning (of characteristic attributes),
6. evaluation and interpretation based on a combination of relevant information and frequencies of mentioning,
7. creating meme maps based on co-occurrence of categories and calculating relevant network characteristics (e.g., centralities).

---

[4]Note that the minimum duration of employment was around 2 years and the maximum duration of employment was around 18 years at the time of the interviews.

The interviews covered a range of questions surrounding the topic of P3 automotive's culture.[5] For the sake of brevity and relevance, the analysis in Sect. 4.4.3 concentrates on the answers given to one particular interview question: "How would you describe P3 automotive's organizational culture?" (authors' translation). Each time an interviewee stressed the relevance of a certain cultural aspect or trait by means of verbal emphasis (e.g., "this is very important") or repetition, the (absolute) frequency of mentioning was increased by one, thereby incorporating a simplified weight with the aim of capturing the prevalence of organizational memes.

To present the results in a way that is illustrative and simultaneously able to capture two central aspects of CMS, namely diversity and interconnectedness of memes, a meme map is generated that depicts the relationships between memes in the answers to the above-mentioned question regarding the description of P3 automotive's organizational culture.

As Martin and Woodside explain in the editorial for their *Journal of Business Research* special section: "Memetics offers the potential to better understand trip decision making by surfacing unconscious thoughts that visitors have about destinations" (Martin and Woodside 2017, p. 112). However, unlike the destination meme maps created by Atadil et al. (2017), the meme map in this article reflects relationships among employee's verbalized units of knowledge in different categories related to organizational culture. Yet, in line with Atadil et al. (2017), the meme map presented below reflects the importance as well as the co-occurrence of (e-)memes. More precisely, following Atadil et al. (2017, p. 156), a meme map is generated "based on two criteria: (1) Frequency of memes, (2) Co-occurrence of memes. Co-occurrence locates pairs of words often found together in respondents' answers to a particular question. Based on the frequency of co-occurrence value, a line appears between a pair of memes. The line's width positively associates with the frequency of co-occurrence." It is important to note, however, that in the present study the meme map is not generated on the basis of word pairs but based on co-occurrence of *categories* (which are often represented by whole statements and not just single words) within an interviewee's answer.[6] Technically speaking, meme maps are undirected but weighted networks based on adjacency matrices that were created by the authors from the content analysis and visualized using first and foremost the open source software *Gephi* (Bastian et al. 2009). To measure the importance of particular memes, the so-called *semantic brand score* (SBS) developed by Fronzetti Colladon

---

[5]More precisely, the authors asked questions about the interviewees' personal motivation for working at P3 automotive, about characteristic features, traits, and driving forces of P3 automotive, behavioral characteristics of P3 automotive's employees, the particularities of P3 automotive's organizational culture in general and with reference to the characteristic features and traits, satisfaction with the culture and areas for improvement, challenges for changing or securing the organizational culture over time, and the impact of specific individuals on the organizational culture.

[6]For example, the *freedom* meme may be actualized in the word freedom but also represented in statements like "I got here [to P3 automotive], didn't know anyone, and could just perform ..., we were allowed to simply burgeon" (IP06, authors' translation) or "I can do what I deem right and sensible. There are no committees, and I don't have to explain myself ..." (IP10, authors' translation).

(2018) is adopted. More precisely, the min-max-normalized SBS is calculated for each meme, thus measuring the *prevalence* (frequency), *diversity* (degree), and *connectivity* (betweenness centrality) of individual memes (see Fronzetti Colladon 2018, p. 154).[7] According to its inventor, the SBS

> is a new measure of brand importance calculated on text data, combining methods of social network and semantic analysis. This metric is flexible as it can be used in different contexts and across products, markets and languages (Fronzetti Colladon 2018, p. 150).

The SBS contributes to the analysis of word co-occurrence networks quite generally because it is not only applicable to brands (Fronzetti Colladon 2018). By combining frequencies and centrality measures, the SBS serves as a useful metric to measure importance of nodes in a network that may also help to fathom the diversity of organizational memes in terms of their importance.

4.4.2 The Complex Memetic System of P3 Automotive

The P3 group is an international consulting firm with headquarters in Aachen, Germany, that was founded as a spin-off of the *Fraunhofer Institute for Production Technology (IPT)* in 1996.[8] Apart from its subsidiary P3 automotive (established 2006), which focuses on management consulting, project management, and various other services for the automotive and supplier industry, the P3 group has established several other subsidiaries. In December 2018, the P3 group was represented in seventeen countries.

P3 automotive with its headquarters in Stuttgart, Germany, has been chosen as a case example for this article due to its remarkable and outstanding culture including its (in-)formal structure and its self-conception as a complex network of more or less independently operating consultants (a staff of about 380 people as of January 2017). Interestingly, there seems to be a strong opposition to fiat, rules, or any kind of command structure, for that matter: Some examples of values that are officially communicated as the self-image of P3 automotive and written on the office walls include "leadership through culture instead of rules," "maximal personal freedom," and "drawing knowledge from the network and sharing it" (translated from Schult 2013, p. 65).

Looking at some of the interviewees' descriptions of P3 automotive's culture in more detail reveals how they have internalized this communiqué, and which importance they assign to their company's special culture:

> I think the culture is P3's most important asset (IP04, authors' translation).

---

[7] For an explanation of what prevalence, diversity, and connectivity represent as a theoretical construct, see Fronzetti Colladon (2018, pp. 151–153).

[8] Note that effective January 10, 2019, the P3 group's holding company split into two companies: P3 group GmbH (headquarters in Aachen) and P3 global GmbH (headquarters in Stuttgart).

4 It's More Than Complicated! Using Organizational Memetics to Capture … 79

> We're a company that's shaped by its culture … This is one of our unique selling points, differentiating us … from others (IP02, authors' translation).

Similarly, the following statements support the notion of leadership through culture:

> We don't make bylaws or other rules, we only make culture (IP06, authors' translation).

> You can by all means lead a complex company by culture, which works quite well for us, as we simply talk to each other. This is very important (IP02, authors' translation).

> So, there are hubs where people meet. Usually, there is a coordinator. But they [the intraorganizational networks] emerge and dissolve in a natural way (IP04, authors' translation).

Moreover, the CMS of P3 automotive's organizational culture underwent changes and is frequently adapting:

> Culture may and shall change. We experience change every day (IP02, authors' translation).

> In a nutshell, we're consistently reinventing ourselves … It comes with the system that things are changing again and again (IP08, authors' translation).

Consequently, P3 automotive's culture depends on both internal and external influences as well as on the changing (importance and interaction of) memes within the CMS itself. More generally, the organizational culture seems to have been heavily influenced by particular individuals or social hubs in the company (especially the founders):

> [Reproducing and transmitting culture] is an executive function. … It's useless to sermonize; the only thing that helps is to set an example, to talk to each other, to spend time with each other (IP04, authors' translation).

> The founders are definitely formative [for the culture]. What … [they] say carries weight. Such people carry on the culture … (IP03, authors' translation).

All of the interviewees argue that every site or location has its own distinct subculture, thus pointing to another property of CMS such as aggregation/hierarchy (cf. Sect. 4.2). These cultural differences are mainly attributed to differences in leadership, clients, geographical peculiarities, and the regional cultural milieu. At the same time, the interviewees emphasize the importance of certain core values:

> It's not about being all alike, but concerning the crucial values, we must be equal. … Of course, we're open for suggestions or changes, everybody is free to somewhat shape these for themselves (IP09, authors' translation).

Interestingly, several interviewees (IP04, IP10, IP11, IP13) stress that they regard the culture as the sum of all values and character traits of co-workers since each employee introduces their own memes into the organizational culture, thereby also influencing behavior at the macro-level as culture is regarded as a dynamic construct (the latter is stressed particularly by IP03). According to some of the interviewees (IP04, IP16, and more implicitly IP05 and IP14), the success of P3 automotive's organizational culture (both in terms of business success and coherence) is also due to the company's recruiting, which works as a filter for selecting individuals carrying memes that are compatible with the organizational ones. Therefore, the same interviewees stress the importance of asking value-oriented questions during the recruiting process. In line with the previous quote related to core values, several members of the organization underscore that preserving the cultural "core" is very important, which, according to IP06, can be achieved as long as the organizational framework continues to enable employees to develop and work freely with the cultural core of the organization continuing to work as a "magnet." An important event mentioned as an external shock to the organizational culture was the global financial crisis of 2009, which for the first time in the company's history made lay-offs necessary, which presented a great challenge for the "family-like" (i.e., informal) organizational culture in terms of a loss of trust (emphasized especially by IP13-IP15).

Generally speaking, the CMS of P3 automotive may actually be considered to operate near "the edge of chaos," that is, "at the boundary between frozen order and chaotic wandering" (Kauffman 1993, p. 31), not just because the interviewees themselves mentioned "less chaos" as the most relevant potential for improving P3 automotive's organizational culture (mentioned nine times in total) but also because the existence of said core values may point to stable attractors in the presence of co-evolving and more rapidly changing cultural entities.

Despite some differences, all 16 interviews revealed that P3 automotive has a unique organizational culture that consists of many different memes or even memeplexes. In the words of Weeks and Galunic:

> We cannot look at memes in isolation. When conceptualizing how culture evolves through a process of the variation, selection, and retention of memes, we must explicitly take into account the fact that memes only make sense when we look at their patterns of combination (Weeks and Galunic 2003, p. 1317).

Therefore, in the following analysis in Sect. 4.4.3, the focus is on the diversity (especially in terms of importance) and the interdependence of P3 automotive's organizational memes.

4.4.3 Meme Map and Results

This article puts forth the first targeted attempt to analyze and visualize an organizational culture as a CMS, thus also conducting one of very few empirical studies on organizational memetics overall. Therefore, the results presented in this section should be seen as a proof of feasibility (akin to the article by Lord and Price 2001) intended as a basis for further inquiry as well as an invitation for refinement. Essentially, two central properties of complex systems are measured and visualized here: diversity and interconnectedness of the system's elements (cf. Sect. 4.2). First, the characteristic elements of P3 automotive's organizational culture are specified based on their prevalence. Subsequently, the interconnectedness of these elements is represented as a meme map built from the (aggregated) co-occurrence networks from the interviewees' answers. Table 4.1 lists all 29 memes identified from the interviews

Table 4.1 Meme frequency (#) and (min-max-normalized) SBS of individual memes

Meme	#	SBS	Meme	#	SBS
Freedom	5	0.49	Entrepreneurship	1	0.18
Sum of individuals	5	1	Excessive growth	1	0.06
Fun	4	0.54	Family-like	1	0.06
Openness	3	0.63	Free radicals	1	0
Personal responsibility	3	0.52	General conditions	1	0.18
Acceptance	2	0.20	Individuality	1	0.18
Amicable	2	0.20	Interconnectedness	1	0.06
Customer orientation	2	0.13	Network	1	0.11
Diversity	2	0.25	Outside the box	1	0.06
Flat hierarchy	2	0.28	Professionalism	1	0.11
Performance-driven	2	0.50	Project culture	1	0.06
Alpha leaders	1	0.06	Soft power	1	0.13
Challenging	1	0.09	Trust	1	0.13
Collegiality	1	0.09	Unstructured	1	0.18
Encouraging	1	0.09			

as constituting P3 automotive's organizational culture and shows their respective frequency and their importance measured as the (min-max-normalized) SBS value. The memes are ranked according to their frequency. Looking at these diverse memes already shows that there is no such thing as *the* culture of P3 automotive but different memes that vary in terms of relevance as cultural building blocks. Most notably, the interviewees prominently conceive of *culture as the sum of individuals*. This perception also reflects the diversity of the CMS because each individual is seen as a contributor to the organizational culture. The other two most prevalent memes (in terms of frequency) are *freedom* and *fun*.

As is known from complexity theory, a CMS cannot be described by simply aggregating the individual memes (cf. Sect. 4.2), though. Because the whole CMS is more than just the sum of individual memes, Fig. 4.1 shows the interconnectedness of the memes represented by a meme map built from the co-occurrence networks of the memes as described in Sect. 4.4.1. Arguably, by making use of the visual intelligence of the brain, instead of just identifying the most frequent memes, one can now also observe different kinds of importance, relationships between memes, and even subcultures or memeplexes.

The node size in the network represents a meme's importance as measured by the min-max-normalized SBS, and the thickness of the links represents the number of co-occurrences. Moreover, the color/shading reflects the membership of memes in presumptive memeplexes. Based on a combination of the visualized meme map and prior insights from the interviews, the authors identify different memeplexes, the most salient ones being what could be called an *individuality memeplex*, a *collective memeplex*, and a *cultural influences memeplex*. The individuality memeplex con-

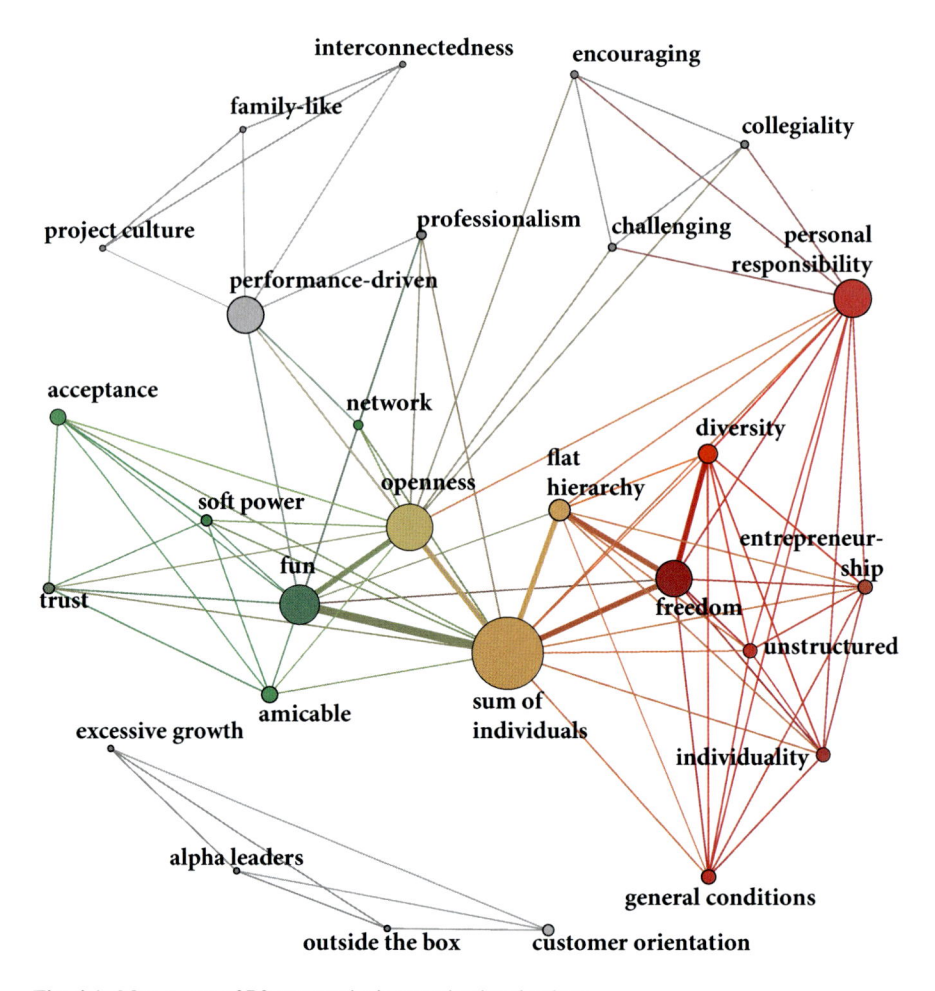

Fig. 4.1 Meme map of P3 automotive's organizational culture

tains the *sum of individuals, openness, freedom, personal responsibility, flat hierarchy, diversity, unstructured, entrepreneurship, individuality*, and *general conditions*.[9] The collective memeplex contains *fun, network, amicable, soft power, acceptance, trust*, and also *flat hierarchy, sum of individuals*, and *openness*. Note that, based on their semantic content, the latter three memes can be considered as parts of both the individuality and the collective memeplex, thus functioning as boundary spanners between the two memeplexes. The cultural influences memeplex is a disconnected component because these memes were verbalized by only one interviewee in response

---

[9] Note that *general conditions* refers to the conditions facilitating self-actualization within P3 automotive.

to the question in focus.[10] The memeplex contains memes that relate to factors with a strong influence on the organizational culture because CMS co-evolve with the environment: *excessive growth* (not just for P3 automotive but as a general characteristic of the culture of the automotive industry as a whole), *customer orientation*, *alpha leaders* (especially founders), and *thinking outside the box*. Although there also exist some memes and connections that elude a straightforward assignment to a coherent memeplex, the meme map in Fig. 4.1 already provides an intelligible representation of the diversity and interconnectedness of P3 automotive's organizational memes. Memes and memeplexes are culturally evolved (and evolving) constructs that should ideally be analyzed over time. Although no longitudinal data on the cultural changes is available for this study, the interviewees have revealed that P3 automotive's culture is frequently changing, while certain core memes appear to exist that act as stable building blocks. These core memes are especially associated with values imprinted by the founders and other charismatic individuals and leaders. Since memeplexes are complex (sub-)systems of mutually compatible, co-adapting memes, the memes that are part of the three above-mentioned memeplexes can be considered to support and influence each other's replication. For example, *freedom* appears to be co-evolving with a *flat hierarchy*, which positively relates with *diversity* but also requires *personal responsibility*. Simultaneously, it is conceivable that *freedom* may sometimes be demanding due to a *lack of structure* but also *fun*, as supported by the following quote:

> But it's simply fun. If you … can also handle the liberties, the partial unstructuredness, the company is great fun [to work at]. But you must be able to cope with this complexity. You have certainly heard of the term 'dynaxity,' which also characterizes us. The term is a combination of dynamics and complexity. It applies quite well to our organization. But, as I said, that's not to everybody's taste (IP02, authors' translation).

In a similar vein, memes within the memeplexes do not only support or influence each other, they can also be considered to even require each other for successful replication. On the one hand, important memes (e.g., freedom) enable the replication of other memes (e.g., diversity) but, on the other hand, necessitate others (e.g., personal responsibility) and may also entail rather undesired ones (e.g., unstructuredness). Consequently, if important memes within a memeplex (e.g., freedom) would be extinguished, this could lead to a cascading failure of the whole memeplex. The fact that some memes appear to be more central and more important (in terms of SBS) within the memeplexes supports the above-mentioned interviewee statements about a cultural core that works as a magnet. In this regard, the most well-connected and most central memes (i.e., flat hierarchy, freedom, fun, diversity, openness, and sum of individuals) may be regarded as core memes within P3 automotive's CMS.

To sum up, the results show how one can reveal the diversity and interdependence of characteristic elements of P3 automotive's organizational culture that can be considered as representations of the underlying organizational memes.

---

[10]The relevance of the cultural influences should not be underestimated, though, because in other parts of the interview (not analyzed here), these cultural influences were mentioned quite frequently also by other interviewees.

4.4.4 Discussion and Limitations

Both the literature review and the results of the exploratory case example suggest that organizational memetics can provide a valuable theoretical and empirical framework for capturing the complexity of an organizational culture. Consistent with the literature on complex systems, complexity here especially means diversity/heterogeneity of components, (non-linear) interconnections, aggregation/hierarchy, adaptation, and emergence. First and foremost, the results shed light on the diversity and interconnectedness of P3 automotive's organizational memes. Moreover, the coherent memeplexes identified by the authors already point to the property of hierarchy in the sense of sub-systems. Although adaptation is not directly captured in the above study (e.g., due to a lack of longitudinal data), the interviews have revealed that P3 automotive's CMS seems to be "dynamically stable" as it appears to have evolved around several relatively resilient core memes while continuously adapting to internal and external influences. Finally, it becomes clear from the meme map that the emergent properties of P3 automotive's CMS—especially in terms of the unwritten rules of behavior which the culture affords—cannot be simply derived from the features of individual memes viewed in isolation. Taken together, the literature review and the case study presented above constitute a rather small but necessary step in the direction of bridging theoretical and applied organizational memetics. In this regard, also some "managerial" implications concerning the preservation of desired memes, potential lock-ins and cultural influences can be derived from this study.[11] When trying to influence or shape an organizational culture, it has to be kept in mind that its CMS emerges from the interconnections and differential fitness of diverse memes. As, for example, also interviewee IP13 notes

> I think of the culture as the sum of all memes, and new memes always come in, some die, it's like genetics. But the general framework must simply be designed in a way that the memes that are good for us can flourish and positively influence each other (IP13, authors' translation).

In this regard, meme mapping has helped to reveal important memes and presumptive memeplexes that may provide some stability in the face of new (and potentially undesirable) memes. In fact, the statement by some interviewees that the "cultural core" can be seen as an important *magnet* supports Price's claim that particular memes may "create a corporate mindset which—without conscious design—acts to preserve the status quo" (Price 1995, p. 306). In the same vein, when mentioning memes and the *stickiness* of certain ideas, Barrett (2015, p. 235) also takes up the metaphor of the magnet: "They [memes] might be informational attractors, like magnets that draw attention." Notice how the notion of attractors also relates back to the literature on complex systems (e.g., see McKelvey 1997, p. 370). However, this also means that one potential managerial challenge for P3 automotive may be a cultural *lock-in*: "once a stable mindset has evolved it seeks to maintain itself,

---

[11] Note, however, that in the face of complexity the term *managerial* does not imply any conscious, top-down ability to design or steer the direction of cultural evolution.

even *in the face of conflicting needs* from the external environment" (Price 1995, p. 306, emphasis added). This effect may even be reinforced due to the company's homophilous hiring, which has also been highlighted by the interviewees (see also Price and Shaw 1998, Chap. 2, and Weeks and Galunic 2003, p. 1332, on a similar issue). By the same token, however, as Waddock (2015, 2016) explains, memes are at the heart of systemic change processes; hence, focusing on particularly attractive or sticky memes may help to avoid unwanted lock-in effects. Put differently, one may be able to (de-)stabilize extant meme(plexe)s but not create an organizational culture by design. Moreover, one possible implication of both the organization's strong focus on the network and the indication that various individuals have a strong influence on the organization's culture is that—although no official hierarchy exists—so-called *prestige hierarchies* may influence the organization's cultural evolution via the heuristics of imitating the successful group members (e.g., Henrich and Gil-White 2001; Stoelhorst and Richerson 2013). Finding out who these "influencers" are within particular intraorganizational networks may, therefore, help to support the stabilization of desired and the change of undesired memes.

Despite these findings, and given the pioneering nature of this endeavor, some limitations must be acknowledged. First, it is important to emphasize again that the findings of the empirical part of this article are not generalizable since the authors only conducted an exploratory study of a single firm. Nevertheless, the case example presented here did not strive for more general implications other than to demonstrate how a memetic perspective can be useful for capturing the complexity and the building blocks of a remarkable organizational culture. Second, having revealed some cultural building blocks does not (yet) yield any information on the variation or transmission of these important memes, let alone *competing* ones. Put differently, despite its apparent tendency to create relatively stable attractors at its core, a CMS evolves, and some of the interviewees have rightfully pointed to the fact that an organizational culture is a *dynamic* system. Hence, the authors acknowledge that they were only able to present a memetic snapshot that neither indicates the direction of the organizational cultural evolution over time nor helps to identify rival or competing memes. Additionally, as the qualitative content analysis relied on subjective interpretations, one cannot completely rule out misinterpretations by the researchers (as also acknowledged by Shepherd and McKelvey 2009, p. 141). Third, the method applied in the present study only captures tip-of-the-iceberg artifacts (or e-memes). The deeper levels of i-memes (mental representations), espoused beliefs and values, or even basic underlying assumptions (Schein 2017) cannot be unequivocally revealed with the approach used here (see also Jonsen et al. 2015, on a related discussion). In other words, the authors could only measure what was verbalized, which may have been subject to social desirability bias despite the attempt to improve the willingness of interviewees to answer truthfully due to anonymization. Therefore, the applicability of the chosen methods as well as the interpretation and transferability of the results is also somewhat restricted. On a related note, it is quite challenging to make sense of the diversity of meme characteristics (as measured by the SBS) and to interpret meme maps for a static network. Hence, a comparison (either with other cases or over time) could help to generate even more insights into the complexity of organizational culture by reaping the full benefits of social network analysis (see, e.g., Fronzetti Colladon's 2018 application of the SBS to investigate brand transitioning over time, or the analysis in combination with an enterprise intranet social network by Fronzetti Colladon and Scettri 2019).

4.5 Conclusion and Avenues for Future Research

Although business researchers are increasingly aware of the complexity of organizational culture in the sense of a CMS, research on organizational memetics is still in its infancy. While the first attempt to combine complexity theory and meme theory to make sense of organizational evolution already dates back to Price's (1995) article almost 25 years ago, conceptual and especially empirical studies are scarce and dispersed as revealed by the above literature review (Sect. 4.3). Ultimately, the review points to the fact that the topic abounds with much potential for further work. The authors have taken this insight as the starting point for a small case study (Sect. 4.4), which has aimed to explore the question if and how the complexity (in this case the diversity and interdependence of characteristic elements) of an organizational culture can be captured such that these elements can be considered as representations of the organizational memes constituting the CMS. The authors have contributed to an answer by presenting lessons from the German consulting firm P3 automotive. Qualitative data gathered via 16 semi-structured in-depth interviews were analyzed to reveal important items or traits, which have been interpreted as (verbal) representations of the underlying organizational memes. Moreover, a meme map was constructed capturing importance and co-occurrences of memes. In summary, this article makes a case for organizational memetics as a valuable theoretical but also empirical resource for capturing the complexity of organizational culture. More precisely, the article's contribution is twofold: First, it contributes to the literature at the intersection of organizational culture, memetics, and complexity theory by presenting the first targeted review of the organizational memetics literature, especially in terms of a conceptual lens for making sense of organizational CMS. Second, it presents the first (exploratory) empirical study in this direction by drawing upon approaches that are related to memetics (e.g., meme mapping) but not yet applied to the question of how the CMS of an organizational culture can be analyzed.

To further advance the field of organizational memetics, the following research directions may be promising: First and foremost, follow-up research is needed to uncover the full Darwinian process of *variation, selection,* and *retention* or *transmission* of P3 automotive's organizational memes, ideally including the ones that have not been selected to illuminate path dependencies and feedback effects. Researchers should, therefore, strive for longitudinal data and could, for example, also combine the approach presented here with a quasi-experimental approach like the one suggested by Shepherd and McKelvey (2009). Moreover, future empirical studies on the complexity of organizational cultures may utilize a combination of social network analysis and memetic analysis (e.g., meme mapping) of the organizational culture in order to reveal co-evolutionary dynamics between influential actors, the organizational culture, and the structure of the organizational network.[12] However, in this latter regard, data availability as well as privacy regulations and confidentiality may be an issue. More generally, though, organizational memetics should also strive to complement data gathered from questionnaires and interviews with data acquired via other methods, including (but not limited to) ethnography, or narrative, thematic, and discourse analyses of an organization's communications as well as official documents. These approaches are common in qualitative research and can add to a memetic analysis of an organization, thereby contributing to a better understanding of CMS. Potentially, the SBS (Fronzetti Colladon 2018), which has been applied above to capture the importance of memes, may even be used as a predictor of a company's performance (e.g., in terms of the stock price; see Fronzetti Colladon and Scettri 2019). Therefore, future research could explore which predictions can be derived especially from the differential importance of memes (as captured by SBS) within the CMS about the organizational cultural evolution (e.g., in terms of the splitting or merging of different subcultures and subsidiaries). Nevertheless, despite these and other potential avenues for future research, the present article already contributes to the literature on the challenges and opportunities of complexity in business research by providing an up-to-date literature review and a proof of feasibility for using organizational memetics to capture the complexity of organizational culture.

Appendix

See Table 4.2.

---

[12] This may be particularly promising within those organizations where social networks and prestige hierarchies are more important than formal hierarchies, i.e., command structures.

Table 4.2 Key contributions to the field of organizational memetics

Author(s)	Approach/method	Goal	Definition or use of memes	Key findings/contribution
Price (1995)	Theoretical/conceptual, literature-based	To argue that "organizational evolution (learning) can be considered as a selection process between mental replicators" (p. 299).	A meme is regarded as "a composite mindset, a paradigm, or a mental model" (p. 307).	One of the first attempts to utilize memetics in an organizational context.
Speel (1997)	Theoretical/conceptual	To propose a memetic framework for studying and describing policy making in organizations.	The meme is defined as a replicator, which is "a piece of data that is copied from retention system to retention system without too much alteration" (no pagination).	Development of a framework with (mainly theoretical) implications for the study of policy making as well as memetics.
Lord and Price (2001)	Empirical; "memetic similarity analysis" by means of a "phenetic analysis of memetic data" (no pagination)	To report "a successful phylogenic reconstruction of the known pattern of decent of the main post-reformation Christian Churches" (no pagination).	"The organisational replicator, whatever it is, can be termed meme by reason of the specific origin of the term in the context of a general theory of complex organisation enabled by replicators" (no pagination). Operationalized as "putative meme strips" rendered as binary data, i.e., *taxon*.	One of the first empirical approaches to organizational memetics by means of a cladogram derived from a memetic code.

(continued)

4 It's More Than Complicated! Using Organizational Memetics to Capture …

Table 4.2 (continued)

Author(s)	Approach/method	Goal	Definition or use of memes	Key findings/contribution
Vos and Kelleher (2001)	Theoretical/literature-based	To diagnose problems within current merger & acquisition (M & A) theories and provide an alternative theory of corporate behavior.	Memes are considered to be "selfish independent ideas operating only to get themselves copied" (no pagination).	The authors develop a memetic approach to M & A.
Pech (2003)	Conceptual, case examples	(Implicitly) raising awareness for the importance of (organizational) meme management.	Memes "represent the knowledge, views, perceptions, and beliefs communicated from person to person" (p. 111).	The author discusses the relevance of memes and their implications for business organizations based on selected cases (e.g., Rip Curl).
Weeks and Galunic (2003)	Theoretical/literature-based	To propose a theory of the cultural evolution of the firm.	The term meme is used "to refer collectively to cultural modes of thought (ideas, beliefs, assumptions, values, interpretative schema, and know-how)" (p. 1309).	The authors provide a new perspective (i.e., a meme's eye view) on "why we have the firms that we have" (p. 1309). They conceptualize the firm as a culture-bearing entity; some elements of culture enhance the performance of the organization while others do not.

(continued)

Table 4.2 (continued)

Author(s)	Approach/method	Goal	Definition or use of memes	Key findings/contribution
Pech and Slade (2004)	Theoretical/conceptual, case examples	Propose a framework that may allow managers to "develop a heuristic for diagnosis of memes and their impact upon organisational culture and execution of the mission" (p. 452).	Employee mindsets that are "conveyed in messages or packets of information that can be termed memes" (p. 452). "Memes are potentially powerful, self-replicating, and sometimes dangerous packets of information that move through copycat behaviours amongst willing or vulnerable hosts susceptible to their message" (p. 463).	Development/proposition of two frameworks ("memetic mapping" of meme fitness, resistance to change, and of toxicity level in the organization) as tools for "memetic engineering" in organizations.
Taylor and Giroux (2005)	Theoretical/conceptual	The authors contribute to organizational ecology literature by proposing a memetically-inspired "theory of organizational retention" (p. 134).	Organizational meme: "a basic replicator that stores the code of an organization and is operative in determining its eventual realized structure" (p. 134).	Proposing a memetically-informed model of organization that illuminates "how an organization might be encoded in the structures of language" (p. 161).
Voelpel et al. (2005)	Conceptual, literature-based with case example(s)	To propose a tool that can help to identify and leverage "the replicators of an organizational innovation culture" (p. 57).	The *organizational meme* is defined as: "Any of the core elements of organizational culture, like basic assumptions, norms, standards, and symbolic systems that can be transferred by imitation from one human mind to the next" (p. 60). The *innovation meme* is defined as: "A unit of cultural transmission or imitation that carries information responsible for innovations, and that can be transferred to other carriers, e.g... employees, departments, organizations etc." (p. 64).	Development/proposition of the innovation meme process as a first approach to meme management in the knowledge economy.

(continued)

Table 4.2 (continued)

Author(s)	Approach/method	Goal	Definition or use of memes	Key findings/contribution
O'Mahoney (2007)	Explanatory/applied memetics: discussion based on a meme's-eye view on qualitative evidence from two case studies of business process re-engineering	The author "applies a theory of memetics to help explain the diffusion of management innovations as a dynamic evolutionary process" (p. 1324) involving the algorithmic components of variation, selection, and replication.	Management innovations are understood as "memes that 'infect' organizations, residing in documents and memories (both biological and digital), being transmitted through consultants, training seminars and networks, and by mutating both the innovation itself and its environment to improve the likelihood of its reproduction" (p. 1325).	The case-based discussion "suggests that the prime-movers in innovation diffusion are not just humans but also memes themselves" (p. 1344).
Rosen et al. (2007)	Theoretical/conceptual, literature-based	To generate insights into the development of problem-solving skills in teams by drawing upon and applying the concept of memes.	The meme is defined as "the smallest unit of cultural information that is subject to the process of selection" (pp. 133–134).	The authors conceptualize cultural influences on problem-solving skills of teams in organizations and show how memetic processes of inheritance, mutation, and selection can be used to explain how teams evolve and adapt.
Shepherd and McKelvey (2009)	Empirical, exploratory quasi-experimental design	To explore "whether understanding organizational memetic variation is empirically possible" (p. 135).	Memes are defined as "independent knowledge-based units of meaning that can be (socially) exchanged—*transmitted*—with more or less accurate transfer with or without alteration of meaning" (p. 138).	Key contributions: The authors (1) develop an empirical method to identify and evaluate memes and memetic variation, (2) show differences in variation creation between and within two different settings (project management and Internet), (3) and assess reliability by comparing two settings and three cases (p. 160).

(continued)

Table 4.2 (continued)

Author(s)	Approach/method	Goal	Definition or use of memes	Key findings/contribution
Lord (2012)	Conceptual/meta-methodological	(1) To review the explanatory power of memetics, (2) to argue that memetics has been retarded by philosophical accusations and premature demands, (3) and to speculate about practical applications and consider resuming methodological research initiatives drawing from biology (p. 349).	"The meme can be thought of as a specific type of idea: one with the capacity for copying itself from mind to mind; from person to person, thereby multiplying its presence among a population" (p. 351).	Proposition of a cultural analogue to *Linnæan Systematics*.
Price (2012b)	Conceptual	To conceptualize management fashions as narrative elements competing for replication and resources.	The meme is used as "an umbrella term for the category containing all cultural modes of thought" (p. 339).	The article supports the point of view that it is possible to regard organizations as ecologies of memes (as well as signifiers, narratives, representations, or discourses) (p. 337).
Gill (2013, 2014)	Conceptual, methodological, and empirical (case study: ethnographic participant observation/narrative analysis and graphical representation in *punnett square*)	Gill aims "to evaluate the meme's conceptual basis" (2014, p. 10) and addresses three research questions: "1. Can organisational culture be divided into units? 2. If so, can such units be seen to selfishly replicate? 3. Can an extra-memetic method be devised with which to answer question 1 and 2?" (ibid.).	Memes are conceptualized as optimon-type units or narrative units.	Gill's findings support "the notion that cultural information occurred as a result of evolved biological traits and that it resides in people's brains" (2014, p. 21); the findings do not (necessarily) support the reduction of cultural aspects to distinct neural structures.

(continued)

Table 4.2 (continued)

Author(s)	Approach/method	Goal	Definition or use of memes	Key findings/contribution
Cook (2015)	Conceptual and two small case studies of two departments at a New Zealand polytechnic	To propose that, by using memetics, the complex dynamics of how individuals and organizations cope with change can be better understood. The case studies are used "to illustrate the usefulness … and the value of this approach for organisations which need to implement change effectively" (p. 230).	Cook develops a "meme complexes architecture" that distinguishes between nMemes (natural memes), vMemes (value memeplexes), cMemes (catalyst change memeplexes), and tMemes (technology memeplexes). The author specifies that "while memes might mimic genes in a virus-like replication and subsequently influence biochemistry, they are not biological entities, but instead social and cultural constructions" (p. 233, emphasis removed).	The author demonstrates how memetics can generate insights into individual and organizational ability to handle change. Although empirical findings are limited to the two cases, the contribution demonstrates the practical value of applying organizational memetics.
Swailes (2016)	Conceptual/applied memetics (memetic perspective on the evolution of talent management in organizations)	The author uses a memetic perspective to help explain "why some organizational cultures adapt to work with a talent mind-set" (p. 341).	"Memes can be thought of as social cultural phenomena such as ideas or fashions that, like genes, adapt, replicate, and survive throughout time and which help to explain cultural transmission" (p. 341).	Offers a memetic perspective on the reasons behind the diffusion and evolution of talent management in organizations.

References

Abatecola, G. (2014). Research in organizational evolution. What comes next? *European Management Journal, 32*(3), 434–443. https://doi.org/10.1016/j.emj.2013.07.008.

Abatecola, G., Belussi, F., Breslin, D., & Filatotchev, I. (2016). Darwinism, organizational evolution and survival: Key challenges for future research. *Journal of Management and Governance, 20*(1), 1–17. https://doi.org/10.1007/s10997-015-9310-8.

Adamic, L. A., Lento, T. M., Adar, E., & Ng, P. C. (2016). Information evolution in social networks. In P. N. Bennett, V. Josifovski, J. Neville, & F. Radlinski (Eds.), *Proceedings of the Ninth ACM International Conference on Web Search and Data Mining - WSDM'16* (pp. 473–482). New York: ACM Press.

Aldrich, H. E., & Ruef, M. (2006). *Organizations evolving* (2nd ed.). London: Sage.

Allen, P., Maguire, S., & McKelvey, B. (Eds.). (2011). *The SAGE handbook of complexity and management*. London: Sage.

Arthur, W. B. (2015). *Complexity and the economy*. Oxford: Oxford University Press.

Atadil, H. A., Sirakaya-Turk, E., Baloglu, S., & Kirillova, K. (2017). Destination neurogenetics: Creation of destination meme maps of tourists. *Journal of Business Research, 74*, 154–161. https://doi.org/10.1016/j.jbusres.2016.10.028.

Aunger, R. (Ed.). (2000). *Darwinizing culture: The status of memetics as a science*. Oxford: Oxford University Press.

Aunger, R. (2002). *The electric meme: A new theory about how we think*. New York: Free Press.

Balkin, J. M. (1998). *Cultural software: A theory of ideology*. New Haven, CT: Yale University Press.

Bandte, H. (2007). *Komplexität in Organisationen: Organisationstheoretische Betrachtungen und agentenbasierte Simulation*. Wiesbaden: Deutscher Universitäts-Verlag.

Bar-Yam, Y. (2004). *Making things work: Solving complex problems in a complex world*. Cambridge, MA: Knowledge Press.

Barrett, H. C. (2015). *The shape of thought: How mental adaptations evolve*. Oxford: Oxford University Press.

Bastian, M., Heymann, S., & Jacomy, M. (2009). Gephi: An open source software for exploring and manipulating networks. In *Proceedings of the Third International ICWSM Conference* (pp. 361–362). Palo Alto, CA: Association for the Advancement of Artificial Intelligence.

Baum, J. A. C., & Singh, J. V. (Eds.). (1994). *Evolutionary dynamics of organizations*. New York: Oxford University Press.

Beinhocker, E. D. (2006). *The origin of wealth: Evolution, complexity, and the radical remaking of economics*. Boston, MA: Harvard Business School Press.

Blackmore, S. (1999). *The meme machine*. Oxford: Oxford University Press.

Boulton, J. G., Allen, P. M., & Bowman, C. (2015). *Embracing complexity: Strategic perspectives for an age of turbulence*. Oxford: Oxford University Press.

Breslin, D. (2014). Calm in the storm: Simulating the management of organizational co-evolution. *Futures, 57*, 62–77. https://doi.org/10.1016/j.futures.2014.02.003.

Breslin, D. (2016). What evolves in organizational co-evolution? *Journal of Management and Governance, 20*(1), 45–67. https://doi.org/10.1007/s10997-014-9302-0.

Burnes, B. (2005). Complexity theories and organizational change. *International Journal of Management Reviews, 7*(2), 73–90. https://doi.org/10.1111/j.1468-2370.2005.00107.x.

Chielens, K., & Heylighen, F. (2005). Operationalization of meme selection criteria: Methodologies to empirically test memetic predictions. In The Society for the Study of Artificial Intelligence and the Simulation of Behaviour (Ed.), *Proceedings of the Joint Symposium on Socially Inspired Computing* (pp. 14–20). Hatfield: University of Hertfordshire. Retrieved from http://www.aisb.org.uk/publications/proceedings/aisb2005/9_Soc_Final.pdf.

Cook, J. E. (2015). Social and cultural influences on organisational change: The practical role of memeplexes. In T. Christensen (Ed.), *Innovative development* (pp. 230–255). Tucson, AZ: Integral Publishers.

Dawkins, R. (1976). *The selfish gene.* Oxford: Oxford University Press.
Dawkins, R. (2015). *Brief candle in the dark: My life in science.* London: Bantam Press.
Dennett, D. C. (1995). *Darwin's dangerous idea: Evolution and the meanings of life.* New York: Simon & Schuster.
Dennett, D. C. (2017). *From bacteria to Bach and back: The evolution of minds.* New York: W.W. Norton & Company.
Dosi, G., & Marengo, L. (2007). Perspective—On the evolutionary and behavioral theories of organizations: A tentative roadmap. *Organization Science, 18*(3), 491–502. https://doi.org/10.1287/orsc.1070.0279.
Edmonds, B. (1999). What is complexity? - The philosophy of complexity per se with application to some examples in evolution. In F. Heylighen, J. Bollen, & A. Riegler (Eds.), *The evolution of complexity* (pp. 1–16). Dordrecht: Kluwer.
Frank, K. A., & Fahrbach, K. (1999). Organization culture as a complex system: Balance and information in models of influence and selection. *Organization Science, 10*(3), 253–277. https://doi.org/10.1287/orsc.10.3.253.
Fronzetti Colladon, A. (2018). The semantic brand score. *Journal of Business Research, 88*, 150–160. https://doi.org/10.1016/j.jbusres.2018.03.026.
Fronzetti Colladon, A., & Scettri, G. (2019). Look inside. Predicting stock prices by analysing an enterprise intranet social network and using word co-occurrence networks. *International Journal of Entrepreneurship and Small Business, 36*(4), 378–391. https://doi.org/10.1504/IJESB.2019.098986.
Gabora, L. (1997). The origin and evolution of culture and creativity. *Journal of Memetics - Evolutionary Models of Information Transmission, 1*. Retrieved from http://cfpm.org/jom-emit/1997/vol1/gabora_l.html.
Galunic, D. C., & Weeks, J. R. (2005). Intraorganizational ecology. In J. A. C. Baum (Ed.), *The Blackwell companion to organizations* (pp. 75–97). Malden, MA: Blackwell.
Gill, J. (2012). An extra–memetic empirical methodology to accompany theoretical memetics. *International Journal of Organizational Analysis, 20*(3), 323–336. https://doi.org/10.1108/19348831211243839.
Gill, J. (2013). *Evaluating memetics: A case of competing perspectives at an SME* (Unpublished doctoral dissertation, Sheffield Hallam University, Sheffield).
Gill, J. (2014). Evaluating the meme concept: The case for a cultural optimon. In *Paper Presented at the EURAM Annual Conference, 4–7 June 2014, Valencia.* Retrieved from http://shura.shu.ac.uk/8171/3/Gill_Evaluating_the_Meme_Concept.pdf.
Hamlin, R. P., Bishop, D., & Mather, D. W. (2015). 'Marketing earthquakes': A process of brand and market evolution by punctuated equilibrium. *Marketing Theory, 15*(3), 299–320. https://doi.org/10.1177/1470593115572668.
Hazy, J. K., Goldstein, J. A., & Lichtenstein, B. B. (Eds.). (2007). *Complex systems leadership theory: New perspectives from complexity science on social and organizational effectiveness.* Mansfield, MA: ISCE Publishing.
Henrich, J., & Gil-White, F. J. (2001). The evolution of prestige: Freely conferred deference as a mechanism for enhancing the benefits of cultural transmission. *Evolution and Human Behavior, 22*(3), 165–196. https://doi.org/10.1016/S1090-5138(00)00071-4.
Heylighen, F., & Chielens, K. (2009). Evolution of culture, memetics. In R. A. Meyers (Ed.), *Encyclopedia of complexity and systems science* (pp. 3205–3220). New York: Springer.
Hofstede, G. (1998). Attitudes, values and organizational culture: Disentangling the concepts. *Organization Studies, 19*(3), 477–493. https://doi.org/10.1177/017084069801900305.
Holland, J. H. (2014). *Complexity: A very short introduction.* Oxford: Oxford University Press.
Hull, D. L. (2000). Taking memetics seriously: Memetics will be what we make it. In R. Aunger (Ed.), *Darwinizing culture: The status of memetics as a science* (pp. 43–67). Oxford: Oxford University Press.
Jacobs, M. A. (2013). Complexity: Toward an empirical measure. *Technovation, 33*(4–5), 111–118. https://doi.org/10.1016/j.technovation.2013.01.001.

Johnson, N. F. (2007). *Two's company, three is complexity*. Oxford: Oneworld.
Jonsen, K., Galunic, C., Weeks, J., & Braga, T. (2015). Evaluating espoused values: Does articulating values pay off? *European Management Journal, 33*(5), 332–340. https://doi.org/10.1016/j.emj.2015.03.005.
Kauffman, S. A. (1993). *The origins of order: Self-organization and selection in evolution*. New York: Oxford University Press.
Kauffman, S. A. (1995). *At home in the universe: The search for laws of self-organization and complexity*. New York: Oxford University Press.
Kirman, A. P. (2011). *Complex economics: Individual and collective rationality*. The Graz Schumpeter lectures. London: Routledge.
Kuckartz, U. (2014). *Qualitative text analysis: A guide to methods, practice and using software*. Los Angeles: Sage.
Kuckartz, U. (2019). Qualitative text analysis: A systematic approach. In G. Kaiser & N. C. Presmeg (Eds.), *Compendium for early career researchers in mathematics education* (pp. 181–197). Cham, Switzerland: Springer.
Kumbartzki, J. (2002). *Die interne Evolution von Organisationen: Evolutionstheoretischer Ansatz zur Erklärung organisationalen Wandels*. Wiesbaden: Springer.
Lissack, M. R. (1999). Complexity and management: It is more than jargon. In M. R. Lissack & H. P. Gunz (Eds.), *Managing complexity in organizations* (pp. 11–28). Westport, CT: Quorum Books.
Lord, A. S. (2012). Reviving organisational memetics through cultural linnæanism. *International Journal of Organizational Analysis, 20*(3), 349–370. https://doi.org/10.1108/19348831211254143.
Lord, A. S., & Price, I. (2001). Reconstruction of organisational phylogeny from memetic similarity analysis: Proof of feasibility. *Journal of Memetics - Evolutionary Models of Information Transmission, 5*. Retrieved from http://cfpm.org/jom-emit/2001/vol5/lord_a&price_i.html.
Malik, F. (2016). *Strategy for managing complex systems: A contribution to management cybernetics for evolutionary systems*. Frankfurt-on-Main: Campus Verlag.
Manikandan, K. S. (2009). Memes in organization studies: A preliminary research agenda. Working paper no. 279, Indian Institute of Management Bangalore. https://doi.org/10.2139/ssrn.2140674.
Marsden, P. (1998). Memetics: A new paradigm for understanding customer behaviour and influence. *Marketing Intelligence and Planning, 16*(6), 363–368. https://doi.org/10.1108/EUM0000000004541.
Marsden, P. (2002). Brand positioning: Meme's the word. *Marketing Intelligence and Planning, 20*(5), 307–312. https://doi.org/10.1108/02634500210441558.
Martin, D., & Woodside, A. (2017). Learning consumer behavior using marketing anthropology methods. *Journal of Business Research, 74*, 110–112. https://doi.org/10.1016/j.jbusres.2016.10.020.
Mayring, P. (2000). Qualitative content analysis. *Forum: Qualitative Social Research, 1*(2). https://doi.org/10.17169/fqs-1.2.1089.
Mayring, P. (2014). *Qualitative content analysis: Theoretical foundation, basic procedures and software solution*. Klagenfurt: SSOAR. Retrieved from https://nbn-resolving.org/urn:nbn:de:0168-ssoar-395173.
McKelvey, B. (1997). Perspective—Quasi-natural organization science. *Organization Science, 8*(4), 351–380. https://doi.org/10.1287/orsc.8.4.351.
Murmann, J. P., Aldrich, H. E., Levinthal, D., & Winter, S. G. (2003). Evolutionary thought in management and organization theory at the beginning of the new millennium. *Journal of Management Inquiry, 12*(1), 22–40. https://doi.org/10.1177/1056492602250516.
Murray, N., Manrai, A., & Manrai, L. (2014). Memes, memetics and marketing: A state-of-the-art review and a lifecycle model of meme management in advertising. In L. Moutinho, E. Bigné, & A. K. Manrai (Eds.), *The Routledge companion to the future of marketing* (pp. 331–347). London: Routledge.

O'Mahoney, J. (2007). The diffusion of management innovations: The possibilities and limitations of memetics. *Journal of Management Studies, 44*(8), 1324–1348. https://doi.org/10.1111/j.1467-6486.2007.00734.x.

Pech, R. J. (2003). Memetics and innovation: Profit through balanced meme management. *European Journal of Innovation Management, 6*(2), 111–117. https://doi.org/10.1108/14601060310475264.

Pech, R. J., & Slade, B. (2004). Memetic engineering: A framework for organisational diagnosis and development. *Leadership and Organization Development Journal, 25*(5), 452–465. https://doi.org/10.1108/01437730410544764.

Powell, J. H., & Wakeley, T. M. (2003). Evolutionary concepts and business economics. *Journal of Business Research, 56*(2), 153–161. https://doi.org/10.1016/S0148-2963(01)00283-1.

Price, I. (1995). Organizational memetics? Organizational learning as a selection process. *Management Learning, 26*(3), 299–318. https://doi.org/10.1177/1350507695263002.

Price, I. (1999a). Images or reality? Metaphors, memes, and management. In M. R. Lissack & H. P. Gunz (Eds.), *Managing complexity in organizations* (pp. 165–179). Westport, CT: Quorum Books.

Price, I. (1999b). Steps toward the memetic self - a commentary on Rose's paper: Controversies in meme theory. *Journal of Memetics - Evolutionary Models of Information Transmission, 3*. Retrieved from http://cfpm.org/jom-emit/1999/vol3/price_if.html.

Price, I. (2004). Complexity, complicatedness and complexity: A new science behind organizational intervention? *Emergence: Complexity and Organization, 6*, 40–48.

Price, I. (2009). Space to adapt: Workplaces, creative behaviour and organizational memetics. In T. Rickards, M. A. Runco, & S. Moger (Eds.), *The Routledge companion to creativity* (pp. 46–57). London: Routledge.

Price, I. (2012a). Organizational ecologies and declared realities. In K. Alexander & I. Price (Eds.), *Managing organizational ecologies* (pp. 11–22). New York: Routledge.

Price, I. (2012b). The selfish signifier: Meaning, virulence and transmissibility in a management fashion. *International Journal of Organizational Analysis, 20*(3), 337–348. https://doi.org/10.1108/19348831211243848.

Price, I., & Shaw, R. (1998). *Shifting the patterns: Breaching the memetic codes of corporate performance*. Chalford: Management Books 2000.

Ridley, D. (2012). *The literature review* (2nd ed.). London: Sage.

Rosen, M. A., Fiore, S. M., & Salas, E. (2007). Of memes and teams: Exploring the memetics of team problem solving. In D. H. Jonassen (Ed.), *Learning to solve complex scientific problems* (pp. 131–156). New York: Erlbaum.

Russ, H. (2014). *Memes and organisational culture: What is the relationship?* (Unpublished doctoral dissertation, University of Western Sydney, Sydney).

Sack, G. (2015). Culture as a complex system. *Workshop: Complexity Lens*. Raffles City Convention Centre, Singapore. Retrieved from https://glocomnet.com/library/culture-as-a-complex-system-graham-sack.

Schein, E. H. (2017). *Organizational culture and leadership* (5th ed.). Hoboken, NJ: Wiley.

Schlaile, M. P., & Ehrenberger, M. (2016). Complexity, cultural evolution, and the discovery and creation of (social) entrepreneurial opportunities: Exploring a memetic approach. In E. S. C. Berger & A. Kuckertz (Eds.), *Complexity in entrepreneurship, innovation and technology research* (pp. 63–92). Cham: Springer.

Schlaile, M. P., Knausberg, T., Mueller, M., & Zeman, J. (2018a). Viral ice buckets: A memetic perspective on the ALS Ice Bucket Challenge's diffusion. *Cognitive Systems Research, 52*, 947–969. https://doi.org/10.1016/j.cogsys.2018.09.012.

Schlaile, M. P., Zeman, J., & Mueller, M. (2018b). It's a match! Simulating compatibility-based learning in a network of networks. *Journal of Evolutionary Economics, 28*(5), 1111–1150. https://doi.org/10.1007/s00191-018-0579-z.

Schult, F. (2013). *Same same but different*. Aachen, Germany: P3 group/P3 ingenieurgesellschaft mbH.

Shepherd, J. (2002). *An evolutionary (memetic) perspective on 'how and why does organizational knowledge emerge?'* (Unpublished doctoral dissertation, University of Strathclyde, Glasgow).

Shepherd, J., & McKelvey, B. (2009). An empirical investigation of organizational memetic variation. *Journal of Bioeconomics, 11*(2), 135–164. https://doi.org/10.1007/s10818-009-9061-1.

Speel, H.-C. (1997). A memetic analysis of policy making. *Journal of Memetics - Evolutionary Models of Information Transmission, 1*. Retrieved from http://cfpm.org/jom-emit/1997/vol1/speel_h-c.html.

Speel, H.-C. (1999). Memetics: On a conceptual framework for cultural evolution. In F. Heylighen, J. Bollen, & A. Riegler (Eds.), *The evolution of complexity* (pp. 229–254). Dordrecht: Kluwer.

Stevenson, A. (Ed.). (2010). *The Oxford dictionary of English* (3rd ed.). Oxford: Oxford University Press.

Stoelhorst, J. W., & Richerson, P. J. (2013). A naturalistic theory of economic organization. *Journal of Economic Behavior and Organization, 90*, S45–S56. https://doi.org/10.1016/j.jebo.2012.12.012.

Swailes, S. (2016). The cultural evolution of talent management. *Human Resource Development Review, 15*(3), 340–358. https://doi.org/10.1177/1534484316664812.

Taylor, J. R., & Giroux, H. (2005). The role of language in self-organizing. In G. A. Barnett & R. Houston (Eds.), *Advances in self-organizing systems* (pp. 131–167). Cresskill, NJ: Hampton Press.

Voelpel, S. C., Leibold, M., & Streb, C. K. (2005). The innovation meme: Managing innovation replicators for organizational fitness. *Journal of Change Management, 5*(1), 57–69. https://doi.org/10.1080/14697010500036338.

Vos, E., & Kelleher, B. (2001). Mergers and takeovers: A memetic approach. *Journal of Memetics - Evolutionary Models of Information Transmission, 5*. Retrieved from http://cfpm.org/jom-emit/2001/vol5/vos_e&kelleher_b.html.

Vos, E., & Varey, R. J. (2012). Marketing as connecting: The ultimate source of happiness and sustainable well-being. *Social Business, 2*(4), 362–371. https://doi.org/10.1362/204440812X13546197293212.

Waddock, S. (2015). Reflections: Intellectual shamans, sensemaking, and memes in large system change. *Journal of Change Management, 15*(4), 259–273. https://doi.org/10.1080/14697017.2015.1031954.

Waddock, S. (2016). Foundational memes for a new narrative about the role of business in society. *Humanistic Management Journal, 1*(1), 91–105. https://doi.org/10.1007/s41463-016-0012-4.

Waddock, S. (2019). Shaping the shift: Shamanic leadership, memes, and transformation. *Journal of Business Ethics, 155*(4), 931–939. https://doi.org/10.1007/s10551-018-3900-8.

Weeks, J., & Galunic, C. (2003). A theory of the cultural evolution of the firm: The intraorganizational ecology of memes. *Organization Studies, 24*(8), 1309–1352. https://doi.org/10.1177/01708406030248005.

Wilensky, U., & Rand, W. (2015). *An introduction to agent-based modeling: Modeling natural, social, and engineered complex systems with NetLogo*. Cambridge, MA: MIT Press.

Williams, R. (2002). Memetics: A new paradigm for understanding customer behaviour? *Marketing Intelligence and Planning, 20*(3), 162–167. https://doi.org/10.1108/02634500210428012.

Williams, R. (2004). Management fashions and fads. *Management Decision, 42*(6), 769–780. https://doi.org/10.1108/00251740410542339.

Wilson, D. S., & Kirman, A. P. (Eds.). (2016). *Complexity and evolution: Toward a new synthesis for economics*. Cambridge, MA: The MIT Press.

Woodside, A. G. (Ed.). (2017). *The complexity turn: Cultural, management, and marketing applications*. Cham: Springer.

Wu, Y., & Ardley, B. (2007). Brand strategy and brand evolution: Welcome to the world of the meme. *The Marketing Review, 7*(3), 301–310. https://doi.org/10.1362/146934707X230112.

Chapter 5
It's a Match! Simulating Compatibility-based Learning in a Network of Networks

Michael P. Schlaile⊙, Johannes Zeman⊙, and Matthias Mueller⊙

Abstract In this article, we develop a new way to capture knowledge diffusion and assimilation in innovation networks by means of an agent-based simulation model. The model incorporates three essential characteristics of knowledge that have not been covered entirely by previous diffusion models: the network character of knowledge, compatibility of new knowledge with already existing knowledge, and the fact that transmission of knowledge requires some form of attention. We employ a network-of-networks approach, where agents are located within an innovation

---

This chapter has been previously published with open access and should be cited as Schlaile, M.P., Zeman, J., & Mueller, M. (2018). It's a match! Simulating compatibility-based learning in a network of networks. *Journal of Evolutionary Economics, 28*(5), 1111–1150. doi: https://dx.doi.org/10.1007/s00191-018-0579-z. We thank Raul J. Kraus for the many valuable discussions during model development. Furthermore, we thank Ivan Savin, Piergiuseppe Morone, Sander van der Hoog, and the other participants of the 16th International J. A. Schumpeter Society Conference in Montreal (July 2016) as well as the attendees of a doctoral seminar organized by Michael Schramm in Hohenheim (April 2014) for helpful suggestions and comments on earlier drafts. We are also indebted to the two anonymous reviewers who contributed to the evolution of this article and to Christoph Junker for discussions at early stages of model development. Needless to say, all remaining errors and omissions are our own. J. Z. gratefully acknowledges funding from the Deutsche Forschungsgemeinschaft (DFG) through the cluster of excellence Simulation Technology (EXC 310 SimTech), and M. M. gratefully acknowledges financial support from the Dieter Schwarz Stiftung. Note on shared first authorship: Michael P. Schlaile and Johannes Zeman contributed equally.

M. P. Schlaile (✉)
University of Hohenheim, Institute of Economics (520) and Institute of Education, Labor and Society (560), Wollgrasweg 23, 70599 Stuttgart, Germany
e-mail: schlaile@uni-hohenheim.de

J. Zeman
University of Stuttgart, Institute for Computational Physics (ICP), Allmandring 3, Stuttgart 70569, Germany
e-mail: zeman@icp.uni-stuttgart.de

M. Mueller
University of Hohenheim, Institute of Economics (520), Wollgrasweg 23, 70599 Stuttgart, Germany
e-mail: m_mueller@uni-hohenheim.de

© The Author(s), under exclusive license to Springer Nature Switzerland AG 2021
M. P. Schlaile (ed.), *Memetics and Evolutionary Economics*, Economic Complexity and Evolution, https://doi.org/10.1007/978-3-030-59955-3_5

network and each agent itself contains another network composed of knowledge units (KUs). Since social learning is a path-dependent process, in our model, KUs are exchanged among agents and integrated into their respective knowledge networks depending on the received KUs' compatibility with the currently focused ones. Thereby, we are also able to endogenize attributes such as absorptive capacity that have been treated as an exogenous parameter in some of the previous diffusion models. We use our model to simulate and analyze various scenarios, including cases for different degrees of knowledge diversity and cognitive distance among agents as well as knowledge exploitation versus exploration strategies. Here, the model is able to distinguish between two levels of knowledge diversity: heterogeneity within and between agents. Additionally, our simulation results give fresh impetus to debates about the interplay of innovation network structure and knowledge diffusion. In summary, our article proposes a novel way of modeling knowledge diffusion, thereby contributing to an advancement of the economics of innovation and knowledge.

5.1 Introduction

Knowledge is important for the economic system both as input and output (see, e.g., Ancori et al. 2000; Antonelli and Link 2015; Barley et al. 2018; Foray 2004; Mokyr 2002, 2017; Smith 2000) and a central building block of innovation and economic evolution (e.g., Audretsch and Feldman 1996; Dosi 1988; Jensen et al. 2007; Lundvall 2004, 2016; Lundvall and Johnson 1996). Consequently, many authors study knowledge dynamics as a fundamental pillar of innovation both from empirical and theoretical perspectives (see, e.g., Morone and Taylor 2010, for a review). Recently, several agent-based models and simulations have emerged that aim to capture various dynamics of knowledge creation and diffusion in (innovation) networks (e.g., Ahrweiler et al. 2016; Bogner et al. 2018; Cowan and Jonard 2004; Gilbert et al. 2007, 2014; Luo et al. 2015; Morone and Taylor 2004, 2010; Mueller et al. 2017; Schmid 2015; Tur and Azagra-Caro 2018; Vermeulen and Pyka 2017). Knowledge and information as well as their diffusion can be modeled in various forms (see, e.g., Cowan and Jonard 2004; Ferrari et al. 2009; Morone et al. 2007; Weng 2014, for different approaches). In previous models, knowledge has often been represented as a vector of knowledge types or categories (e.g., Cowan and Jonard 2004; Luo et al. 2015; Mueller et al. 2017). However, as Piergiuseppe Morone and Richard Taylor (2010, p. 37) note: "considering knowledge as a number (or a vector of numbers) … restricts our understanding of the complex structure of knowledge generation and diffusion." Arguably, when representing knowledge itself as an easily quantifiable cumulative entity in terms of numbers or vectors, the analysis of knowledge diffusion processes may provide an incomplete picture. There are many examples where knowledge generation and diffusion involve more than "stockpiling" additional pieces of knowledge, for example, by establishing meaningful connections and complex relations (cf. *relational knowledge*, Halford et al. 2010). Consequently, knowledge is not only cumulative (e.g., Boschma 2005; Foray and Mairesse 2002), it can also be tacit or sticky (see, e.g., Antonelli 2006; Cowan et al. 2000; Polanyi 1966; Szulanski 2003; von Hippel 1994), and learning must be considered as a path-dependent process (e.g., Baddeley 2010, 2013; Boschma and Lambooy 1999; Dosi et al. 2001; Rizzello 2004; Hayek 1952).

The aim of our article is to contribute to this broad line of research by proposing a *meso-analytic*[1] agent-based simulation model that presents an alternative approach to modeling knowledge and its diffusion by taking into account the relational, cumulative, and path-dependent aspects of (social) learning. More specifically, we explicitly take into account the following inherent characteristics of knowledge:

- The network character of knowledge,
- compatibility of newly acquired with already existing knowledge (an important aspect of path dependence in social learning processes), and
- competition among knowledge units for attention.

Consequently, the central purpose of our article is to shed light on the potential implications and advantages of modeling knowledge and its diffusion differently. We develop a model that can more adequately capture the complexity of diffusion processes resulting from incorporating these three characteristics of knowledge. At the same time, the model is kept simple enough to be analyzed in a conclusive manner.

The article is structured as follows: Section 5.2 presents the theoretical background of our approach. In particular, we review important literature and elucidate the article's research focus (Sect. 5.2.1), motivation and foundations (Sect. 5.2.2), and then describe our model (Sect. 5.2.3). Subsequently, in Sect. 5.3, we present the analyses of the simulation results. More precisely, in Sect. 5.3.1, we start with the results of a baseline scenario that aims to illuminate the changing diffusion dynamics once we explicitly consider the network character of knowledge- and compatibility-based learning. Building on this, Sect. 5.3.2 tackles the important topic of knowledge diversity within and between agents in an innovation network and the consequences for the performance of knowledge exploitation versus exploration strategies. Thereafter, Sects. 5.3.3 and 5.3.4 address the issue of measuring knowledge diffusion—while taking the three characteristics mentioned above seriously—both on aggregated (Sect. 5.3.3) and individual (Sect. 5.3.4) levels. Finally, Sect. 5.3.5 contributes to discussions on the impact of an innovation network's structural properties on the overall diffusion performance. In the last section of our article (5.4), we draw our conclusion and suggest directions for future work.

---

[1] For more information on the relevance of meso-economic analysis, see Dopfer (2012), Dopfer et al. (2004), or Dopfer and Potts (2008, Chap. 4).

5.2 Theoretical Background

5.2.1 Relevant Literature and Research Focus

Diffusion research can take various forms. It ranges from abstract and purely theoretical contributions to applied studies analyzing the effects of specific diffusion processes in detail. Any attempt to present a comprehensive overview would clearly go beyond the scope of this article.[2] On a general note, we can, however, identify and differentiate four focus areas in the diffusion literature that address different questions, which are also important for this article's focus on diffusion of knowledge:

(a) *what* diffuses?
(b) *how* does it diffuse?
(c) *where* does it diffuse?
(d) what are the *effects* (or performance) of the diffusion process and *how are they measured*?

To give just a brief summary, with regard to level "(a) what diffuses," method and model range from capturing knowledge as simple information (e.g., numbers) or virus-like entities to more complex representations in terms of vectors, bit strings, or even graphs (see, e.g., Morone and Taylor 2010, for a review). The next important question is then, "how does it diffuse?" (level b). This, of course, is to a certain extent also dependent on the representation of knowledge. Sometimes, diffusion exhibits features of *simple contagions* (such as infectious diseases), whereas on other occasions, one can observe *complex contagions*, which are affected by *homophily* and *social reinforcement* (Tur et al. 2014, 2018; Weng 2014; Weng et al. 2013; see also Lerman 2016). Moreover, this level concerns the way the diffusion process itself is designed. In some models, knowledge is exchanged via a barter trade mechanism, meaning that knowledge will only be exchanged if all partners involved can somehow benefit from the exchange (e.g., Cowan and Jonard 2004), whereas in others, knowledge may flow freely (see also Klarl 2014; Morone and Taylor 2009, on these different diffusion or transfer mechanisms). In addition, many of the models with an economic focus also incorporate the *absorptive capacity*[3] of firms, which also influences the diffusion process (e.g., Cowan and Jonard 2004; Egbetokun and Savin 2014; Savin and Egbetokun 2016). For level "c) where does it diffuse?", the modeling approaches to depict the underlying (social) structure range from diffusion or exchange on a grid or *von Neumann neighborhood* to complex network architectures (e.g., random, small-world, and scale-free). Sometimes, models are also combined with real-world network data (e.g., Bogner et al. 2018) or with dynamic or (co-)evolving network structures (e.g., Luo et al. 2015; Tur and Azagra-Caro 2018).

---

[2]For rather general reviews of diffusion literature, we refer the reader to Jackson and Yariv (2011), Lamberson (2016), and Valente (2006).

[3]According to Wesley Cohen and Daniel Levinthal (1990, p. 128), an individual's or firm's absorptive capacity is the ability to evaluate and utilize external knowledge, which "is largely a function of the level of prior related knowledge."

Alternatively, researchers have focused on the spatial dimension of knowledge diffusion (e.g., Canals et al. 2008). Finally, concerning the last level (d), the underlying question is how diffusion and its performance as well as effects are *measured*. This also differs between the various approaches, depending on the modelers' focus. First, based on how knowledge is represented in level a, we could measure how much and how fast knowledge diffuses. This can be done either on an individual, i.e., *micro*-level, or on a systemic *macro*-level. Second, some models also shift the focus and measure "indirect" or economic effects of knowledge diffusion, e.g., in terms of increased returns, more product innovations, etc. (e.g., Cowan and Jonard 2007).

It is important to note, however, that these four levels can only to some extent be regarded in isolation. In fact, many models incorporate and analyze possible *feedback effects*. This means, for example, that—depending on how knowledge diffuses (level b) and on the effects of that process (level d)—other levels may be affected as well. In other words, the mechanisms and results of previous diffusion processes can also influence future performance either by changing an individual agent's influence or importance (e.g., by becoming an *opinion leader* or a *gatekeeper*) or, at the systemic level, by dynamically changing the underlying social structure (level c). If the latter is represented as a network, diffusion impinges upon the network topology. This occurs, for instance, via link formation or deletion between the agents in the network, which in turn affects future diffusion (see, e.g., Canals 2005; Luo et al. 2015; Weng 2014; Tur and Azagra-Caro 2018).[4]

Nevertheless, at this point, our focus lies on the representation of knowledge (level a). As already mentioned above, previous approaches have modeled knowledge in different ways, all of which have particular merits and limits. Probably the simplest way to model knowledge is to assign a scalar number to each agent that represents the total amount of knowledge possessed by that agent. However, with such a simple approach it is impossible to differentiate qualitatively *what kind* of knowledge or information an agent holds. One way to tackle this problem is to model an agent's knowledge as a vector, where each entry in the vector represents the "knowledge stock" of an agent in a particular knowledge *category*. Whenever an agent receives new knowledge related to a specific category, the respective number in the knowledge vector is increased (see, e.g., Bogner et al. 2018; Cowan and Jonard 2004; Luo et al. 2015; Mueller et al. 2017, for models using this knowledge representation). This approach already allows for a more qualitative view of knowledge while preserving its quantitative aspect, as the total knowledge is simply the sum of all entries in the vector. Nevertheless, it is not a priori clear if and how different knowledge categories are related to each other, and whether newly acquired knowledge can be relevant in several of such categories at the same time. Moreover, when knowledge is transferred to an agent, there is no way of determining whether the specific "kind" of received knowledge is already known to the agent or not. Consequently, there is no obvious way of representing how agents can learn *new* knowledge from each other, and thus models incorporating such an approach usually assume that knowledge can only flow from those agents with a higher level of knowledge in a specific category to agents

---

[4]For a general review of coevolutionary network dynamics, see Gross and Blasius (2008).

with a lower level in that category. Put differently, since numbers or vectors imply a rather easy quantifiability, they might obfuscate the complex nature of knowledge and learning, which often involves the creation of meaningful *connections* between ideas and concepts as well as their recombination (see also Arthur 2007; Markey-Towler 2016, 2017; Tywoniak 2007; Vermeulen and Pyka 2017, for related discussions). As Bernard Ancori and his coauthors explained almost two decades ago, "knowledge is not a mere stock resulting from the accumulation of an information flux" (Ancori et al. 2000, p. 259). In order to overcome this shortcoming, a representation of knowledge is required which ensures that what is learned is uniquely identifiable. One possibility to achieve this is a quantization of knowledge into distinct *units*, where each knowledge unit (KU) has a unique signature. The knowledge an agent possesses is then represented by a set of KUs. Still, a simple set of KUs provides no notion of how these units may be related to each other. As Bart Nooteboom argues, one important implication of the approaches of *connectionism* and *parallel distributed processing* "is that knowledge is not stored in units, to be retrieved from there, but in patterns of activation *in connections between units*" (Nooteboom 2009, p. 51, emphasis added). We have, for this and other reasons explained below, decided to follow Paolo Saviotti (2009, 2011), who asserts that it is "possible to represent knowledge as a *network*" (Saviotti 2011, p. 151, italics in original). Therefore, in this work, we go one step further by connecting individual KUs based on a pairwise relation. Thus, the total knowledge an agent holds becomes a *network* of KUs.

5.2.2 Motivation and Foundations

Although we approach this complex endeavor by means of an abstract computational model, this article is motivated by—and rests upon—the theoretical background of *innovation systems* (see, e.g., Klein and Sauer 2016, for a review) and particularly *innovation networks* (e.g., Ahrweiler and Keane 2013; Buchmann and Pyka 2012; Koschatzky et al. 2001; Pyka and Küppers 2002). More precisely, we adopt the view of Tobias Buchmann and Andreas Pyka that the social structures where knowledge diffuses (our above level c) can be represented as innovation networks that "consist of actors and linkages among these actors. The idea of actors is conceived very broadly and also encompasses besides firms, individuals, research institutes and university laboratories, venture capital firms or even standardization agencies. Links among the actors are used as channels for knowledge and information flows ..." (Buchmann and Pyka 2012, p. 467).[5]

---

[5]For the purpose of our article, we follow Everett Rogers' notion of an *innovation* as "an idea, practice, or object that is perceived as new by an individual or other unit of adoption" (Rogers 2003, p. 12). One may be tempted to reduce knowledge to the "idea part" of this definition, but—even on a more general note—Dorothy Leonard argues, for instance, that both generation and diffusion of innovation(s) can be viewed "as a process of knowledge generation and transfer" (Leonard 2006, p. 85) and that, regardless of their outward form or medium, innovations can be "considered to be essentially bundles of *knowledge*" (ibid., p. 86, italics in original).

Since a novel way of representing knowledge (level a) also implies changes in how models capture its diffusion (level b), we require further theoretical foundations. Therefore, our approach also builds on the *memetics* literature, among others, especially when *memes* are understood as units of information transmitted primarily via social learning processes (e.g., Heylighen and Chielens 2009; von Bülow 2013, for an overview).[6] In this sense, ideas or units of knowledge may also be conceived as memes "made" of (semantic) information (Dennett 1995, 2017) that can diffuse through or on social networks of agents (e.g., Gupta et al. 2016; Spitzberg 2014; Weng 2014).[7] More importantly, the memetics literature also supports the idea of representing knowledge as a network based on the idea of *memeplexes* (e.g., Speel 1999), which may be conceived as complex systems or complex networks of memes that can replicate more successfully in an aggregated or connected form than the isolated memes on their own (Blackmore 1999; Heylighen and Chielens 2009; Schlaile 2018).

This element of interconnection and interdependence also relates to another important aspect of our approach, namely, that ideas, KUs, or memes need to have a certain *compatibility* with already existing ones: "An idea that is more compatible is less uncertain to the potential adopter and fits more closely with the individual's situation. Such compatibility helps the individual give meaning to the new idea so that it is regarded as more familiar" (Rogers 2003, p. 240).[8] In short, "knowledge requires knowledge to be assimilated" (Morone and Taylor 2010, p. 49, with reference to Ancori et al. 2000).

Note that the aspect of compatibility is also closely related to the concept of (optimal) *cognitive distance* (e.g., Boschma 2005; Nooteboom 1999, 2009; Nooteboom et al. 2007; Wuyts et al. 2005), because compatibility does not only mean similarity. The notion of cognitive distance implies that agents can only learn from each other and innovatively utilize the knowledge they exchange if their cognitions are neither too similar nor too different (see also Bogner et al. 2018). According to Nooteboom et al. (2007, p. 1017), "The challenge ... is to find partners at sufficient cognitive distance to tell something new, but not so distant as to preclude mutual understanding." Indeed, this leads us to another aspect that distinguishes our approach from previous ones. Whereas cognitive distance can be conceived as a property at the agent level, compatibility focuses on the *knowledge level*. A network of compatible KUs or memes can thus be regarded as the shared mental representations necessary for con-

---

[6] However, as John Langrish stresses, it may be (more) suitable to regard memes as "patterns of thought" instead of units (Langrish 2017, p. 315).

[7] Note that there are certain reservations (not only) among economists to take up the concept of memes (see, e.g., Hodgson and Knudsen 2010, 2012; Roy 2017), and it can be argued that the model presented below at this stage deliberately neglects some aspects of Darwinian evolution (particularly, extinction and mutation or recombination of these units), so it may not yet be appropriate to call it a genuinely memetic model.

[8] Although we do not confine ourselves to individuals as the units of adoption, we focus specifically on the compatibility of a KU with a potential adopter's prior/existing knowledge, which may also include cultural ideas or "cultural knowledge" held by individuals or organizations (e.g., see Sackmann 1991, on cultural knowledge in organizations).

structing (interpretative) schemata, which in turn serve human agents as information filters (or models *of* the world) as well as heuristics (or models *for* the world) for discovering, creating, and exploiting opportunities (Schlaile and Ehrenberger 2016).

Another related strand of literature has argued that KUs or memes are faced with the "scarce resource" of *attention* (see also Weng et al. 2012).[9] In line with Rogers (2003), Kate Distin explains "Indeed, selection will often depend on a novelty's *compatibility* with the rest of the meme pool. In their bid to gain and retain our *attention*, memes will succeed best if they fit in with facts and skills that we have already absorbed, being influenced particularly by those to which we are greatly attached" (Distin 2005, p. 205, emphasis added). As Herbert Simon famously argued, information consumes the attention of its recipients (Simon 1971), whereby it may be appropriate to speak of competition for attention among the nodes in the knowledge networks.[10] Viewing attention as a scarce resource allows us to take up and integrate some approaches and notions from the *economics of attention* (e.g., Davenport and Beck 2001; Falkinger 2007, 2008). For the remainder of this article, we adopt the definition of attention proposed by Thomas Davenport and John Beck: "*Attention is focused mental engagement on a particular item of information*" (Davenport and Beck 2001, p. 20, italics in original). Note that in the case of organizational agents (as opposed to individuals), it may make sense to interpret the attention (i.e., "focused mental engagement") of a firm or research institute more in terms of their *current focus*, e.g., in research and development (R&D). This aspect can be assumed to influence *how* knowledge diffuses (level b) and, therefore, also has to play an important role in our model below.

In summary, we propose a *network-of-networks* approach to capture knowledge diffusion and assimilation in innovation networks that also allows for the important issues of knowledge compatibility and scarcity of attention explained above.

5.2.3 Model Description

Agent-based modeling (ABM) has proven to be a useful method for simulating complex, dynamic phenomena (see, e.g., Gilbert 2008; Hamill and Gilbert 2016; Müller 2017; Schmid 2015; Wilensky and Rand 2015), including diffusion processes on networks (Garcia 2005; Kiesling et al. 2012; see also Barrat et al. 2008, or Namatame and Cheng 2016, for extensive reviews). As explained above, we employ a network-of-networks approach, meaning that each agent (e.g., firm or individual) a_i in an innovation network $A = \{a_i | i \in [0, N_A)\}$ of size N_A itself contains a network $B_i = \{b_{ij} | j \in [0, N_B)\}$ of size N_B whose nodes $\{b_{ij}\}$ each contain a meme or

---

[9] As, for instance, Matthew Crawford explains "attention ... [can be] treated as a resource—a person has only so much of it" (Crawford 2015, p. 11), and Josef Falkinger (2008) also addresses the problem of "limited attention as a scarce resource in information-rich economies."

[10] As Joel Mokyr declares (in a different but related context), evolutionary systems exhibit "a property of superfecundity ..., that is, there are more entities than can be accommodated, so there must be some selection in the system" (Mokyr 1998, p. 6).

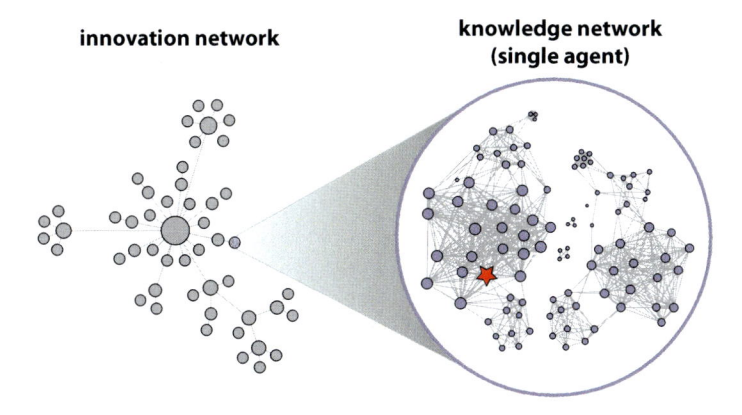

5 It's a Match! Simulating Compatibility-based Learning in a Network of Networks 107

Fig. 5.1 Illustration of the network-of-networks approach. The left side depicts an (arbitrary) innovation network A of agents $\{a_i\}$, whereas the right side shows a magnification of a single agent's knowledge network B_i with nodes $\{b_{ij}\}$ containing one KU K_{ij} per node. The currently focused KU is depicted as a red star.

knowledge unit K_{ij} (see Fig. 5.1).[11] Each K_{ij} is, in turn, represented as a bit string of finite length n_K, where n_K is fixed and identical for all K_{ij} in the model.

Incorporating the aspects of knowledge compatibility and scarcity of attention in our model also affects the way knowledge is exchanged and thereby diffuses through the network of agents. Instead of passively exchanging KUs that just "improve" the knowledge stock of its recipients, new KUs now have to be assimilated and integrated into the existing knowledge network, taking into account that the construction and diffusion of knowledge is a path-dependent process (e.g., Baddeley 2010; Rizzello 2004, with reference to Hayek 1952). As explained in the previous section, successful assimilation of new KUs into the existing knowledge network will be highly dependent on *compatibility* with prior/existing knowledge of the recipient.[12] This aspect represents another important novelty of our model: In several previous models, the absorptive capacity of firms is an exogenous model parameter (e.g., Cowan and Jonard 2004; Mueller et al. 2017), whereas in our model, a firm's ability to recognize the value of new, external knowledge and assimilate it is endogenous, emerging from the compatibility of the received KUs with the agent's existing knowledge network.[13] Our model captures this central aspect by introducing a measure of compatibility c between KUs based on their normalized *Hamming distance d* (Hamming 1950). It

---

[11] Henceforth, for the sake of consistency, we only use the term "knowledge unit" (KU) since the model does not yet capture the whole spectrum of evolutionary dynamics. More specifically, the model does not yet allow for variation and some elements of selection such as forgetting and, hence, differential "fitness" of KUs. Therefore, the term meme may not be fully appropriate at this stage of the model.

[12] In the words of Ray Reagans and Bill McEvily (2003, p. 243), "people learn new ideas ... by associating those ideas with what they already know."

[13] Note that alternative conceptualizations of *endogenous* absorptive capacity have also been proposed, for example, by Klaus Wersching (2010) or Ivan Savin and Abiodun Egbetokun (2016).

is defined as
$$c_{jk} \equiv c(K_{ij}, K_{ik}) = 1 - d(K_{ij}, K_{ik}) \in [0, 1], \qquad (5.1)$$

where the KUs K_{ij} and K_{ik} are bit strings of fixed length n_K. The normalized Hamming distance d can then be expressed as

$$d_{jk} \equiv d(K_{ij}, K_{ik}) = \frac{1}{n_K} \sum_{m=0}^{n_K-1} \left[K_{ij} \oplus K_{ik} \right]_m, \qquad (5.2)$$

where $\oplus$ denotes the logical bitwise XOR ("exclusive or") operator, while the operator $[\cdot]_m$ returns the m-th bit of its operand.

The model is set up as follows: First, the agent network A is filled with $N_A = 100$ agents a_i, and $M_A = 200$ undirected edges are established between pairs of agents according to a user-selected type of network, or loaded from an existing structure. Each of the agents' knowledge networks B_i is then filled with $N_B = 100$ nodes b_{ij} (unless stated otherwise), and each b_{ij} receives a unique KU K_{ij} of length $n_K = 32$ bits. All pairs of nodes (b_{ij}, b_{ik}), $j \neq k$ within an agent a_i's knowledge network B_i are then connected by an undirected edge $e_{i(j,k)}$ if the condition $P_{jk} = 1$ is fulfilled, where P_{jk} is a function of the respective KUs' compatibility:

$$P_{jk} \equiv P(c_{jk}) = \begin{cases} 1 & \text{if } \gamma < c_{jk} < 1 \\ 0 & \text{otherwise} \end{cases}. \qquad (5.3)$$

Each of the edges $\{e_{i(j,k)}\}$ created in this manner is given an edge weight $w_{i(j,k)} = c_{jk}$. The rectangular "compatibility window" described by Eq. (5.3) is chosen solely due to its simplicity and could, in principle, be of any other suitable shape. Nevertheless, the threshold γ has a distinct meaning: We interpret a compatibility of $c_{jk} = 0.5$ as a point of indifference, dividing the compatibility range qualitatively into the lower half with $c_{jk} < 0.5$, where the respective KUs K_{ij} and K_{ik} are believed to be rather incompatible, and the upper half with $0.5 < c_{jk}$, where they are considered to be rather compatible. According to this interpretation, γ should be greater than 0.5. However, the compatibility measure is based on the normalized Hamming distance, which in turn is also a measure of *similarity*. Thus, a compatibility of $c_{jk} = 1$ implies $K_{ij} \equiv K_{ik}$. The condition P_{jk} will also be used to determine the exchange of KUs between agents, and since firms or individuals are usually not interested in learning what they already know, c_{jk} must be strictly smaller than one to fulfill $P_{jk} = 1$. Note that, as already mentioned above, the idea of compatibility is related to the notion of (optimal) cognitive distance. By introducing the condition P_{jk}, we are in line with Nooteboom and others (for an overview, see, e.g., Nooteboom 2009) who argue that learning and innovative performance depend on the agents being able to learn something new while at the same time making sense of the knowledge they exchange. In other words, agents' cognitions should not be too distant from their

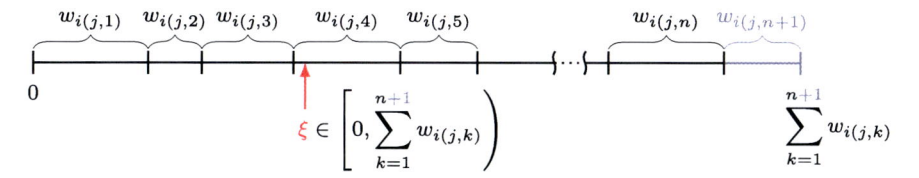

Fig. 5.2 Illustration of the selection mechanism used in the random walk along the edges in a knowledge network B_i. Let b_{ij} be the currently focused node with degree n in the knowledge network B_i. The cumulative sums of the weights of edges connecting b_{ij} with its n neighbors are stored. Additionally, a virtual edge to a randomly chosen node from B_i is considered and its corresponding edge weight is added to the cumulative sums in order to allow random jumps to arbitrary nodes in the network. Then, a random number ξ is uniformly chosen on the interval $\left[0, \sum_{k=1}^{n+1} w_{i(j,k)}\right)$. The index k of the interval to which ξ corresponds is determined, and the random walk is advanced to the thereby chosen k-th neighbor node of b_{ij}.

partners'. By means of our compatibility window, we apply these ideas to the level of the KUs.

Now that the initial setup of the model is complete, we proceed by defining the rules governing its temporal progress. As already explained in the previous section, several scholars have argued that knowledge and information consume the attention of their recipients, resulting in some kind of competition among the KUs for the "scarce resource" of attention (e.g., Simon 1971; Weng 2014; Weng et al. 2012). Moreover, in the context of firms or research institutes, we have proposed that this "attention" could be interpreted as their current *R&D focus*. Our model incorporates shifts of this focus as a weighted random walk from node to node along the existing edges of the agent's knowledge network, starting at a randomly chosen node.

Thus, at every time step t, in ABM often referred to as a *tick*, the focus of each agent a_i changes its position in the agent's knowledge network B_i by advancing from the currently focused node b_{ij} to an adjacent node b_{ik}. Since a node b_{ij} usually has more than one adjacent neighbor node, the probability of advancing to any particular adjacent node b_{ik} is weighted by the connecting edge's weight $w_{i(j,k)}$. The procedure of selecting the next node in the random walk is depicted in Fig. 5.2.

As the weighted random walk depicts the agents' shifting focus in their respective knowledge networks, in the case of organizational agents (e.g., firms), this weighted random walk can also be interpreted as a measure for the agent's strategy of knowledge *exploitation* versus *exploration* (see also the discussion in March 1991; Schmid 2015).[14] For our model, this means that in the latter case, the focused KU would be picked randomly with uniform probability, i.e., without taking the structure of the knowledge network into account. To distinguish the two cases, we will refer to them as (weighted) random walk and (unweighted) random jump, respectively.

Diffusion of knowledge in the innovation network A is realized by communication between pairs of agents according to a simple *knowledge transfer protocol*. For

---

[14]Especially in the context of organizational learning, we may also observe trade-offs between organizational strategies of exploration versus exploitation. We will come back to this in Sect. 5.3.

technical reasons and for the sake of a meaningful analysis by focusing on our knowledge representation as a network (level a), at this stage, we choose not to impose any restrictions on how knowledge diffuses (level b) aside from compatibility.[15] Therefore, the trade protocol chosen at this stage is a *knowledge pull* mechanism with high fidelity (i.e., there are no errors during replication). At every tick (after each step of the random walk), each agent randomly selects *one* of its neighboring agents and retrieves this agent's currently focused KU. The compatibility of this KU with the receiving agent's currently focused one is then evaluated, and, if these two KUs satisfy the condition given in Eq. (5.3), the received KU is integrated into the receiving agent's knowledge network. The integration of such a newly received KU follows the very same procedure as applied during the construction of the initial knowledge network, i.e., edges are established to all previously existing KUs if their compatibility c with the new unit satisfies the condition $P(c) = 1$ according to Eq. (5.3). By choosing this "instantaneous" way of integrating a new KU into a knowledge network, we implicitly make the assumption that the internal communication within each agent is much faster than the external communication between different agents. Further details on the choice of parameters and the technical framework are given in Appendices A and B.

5.3 Results

5.3.1 Baseline Analysis

In order to investigate the effects of compatibility on the dynamics of knowledge diffusion in our model, we first take a step back and eliminate the influence of compatibility completely. This is accomplished by setting the compatibility threshold γ to zero, so that agents can take up any kind of new KUs. Furthermore, we switch off the weighted random walk along the edges in the agents' knowledge networks, which means that an agent's focus can jump to any of its KUs with equal probability. Effectively, this implies that, for a brief moment, we go back to a model where an agent's knowledge is not represented by a network but by a set of unconnected KUs.

If we went back even further and did not consider the *uniqueness* of KUs, i.e., agents could take up KUs they already have, the resulting dynamics would be trivial. Since, then, every agent is allowed to retrieve one KU from a neighboring agent per time step, the number of KUs per agent would simply increase linearly with time. This also means that it would be impossible in this case that the topology of the innovation network A could have any effect on knowledge exchange. In that respect, we want to stress that this simple thought experiment justifies our choice of knowledge trade mechanism, since a simple knowledge pull ensures the elimination of topological effects that are not based on knowledge compatibility.

---

[15]See also Bogner et al. (2018) or the discussion of so-called *knowledge communities*, where trust and reciprocity reinforce the openness of knowledge and norms of knowledge sharing (Foray 2004).

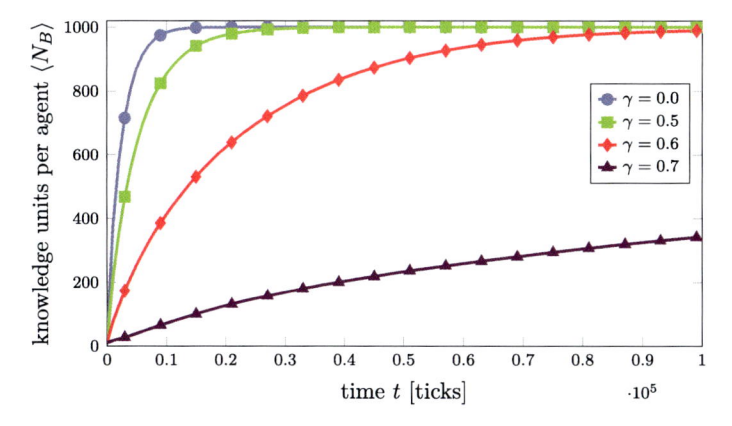

5 It's a Match! Simulating Compatibility-based Learning in a Network of Networks

Fig. 5.3 Average number of KUs $\langle N_B \rangle$ per agent over time in an innovation network A with an Erdős–Rényi (random network) topology for different compatibility thresholds γ. Lines are averages over 100 agents. The standard error of the mean is smaller than the line width and therefore not shown.

Starting from this point, we will now continue by successively incorporating more and more features of our model and study their effects.

First, we only consider the uniqueness of KUs and leave other features of our model turned off. If agents only take up KUs that are new to them, the maximum number of KUs an agent can assimilate is limited to the total number of unique KUs existing in the simulated system, and growth is bounded. Initially, the probability of retrieving a new KU is generally high, and thus growth is fast. However, the more KUs an agent has, the less likely it is for that agent to retrieve new KUs. Therefore, the growth rate decreases over time.

If we now successively increase the compatibility threshold γ while still leaving the random walk switched off (i.e., all agents can be imagined to follow a "knowledge exploration" strategy, where focused attention on any of the agent's KUs is equally likely), the probability to retrieve new knowledge that is *compatible* to an agent's currently focused KU decreases with increasing γ. Figure 5.3 shows the average number $\langle N_B \rangle$ of KUs per agent over time for different values of γ.[16]

The observed effect of an increasing compatibility threshold γ is that the time scale of the simulation is simply stretched, since the growth rate of an agent's knowledge is decreased with increasing γ.[17] When relating this result to a real-world learning process, it trivially means that if it is harder for an agent to find compatible knowledge, learning will take more time.

As a next step, we will establish the actual representation of an agent's knowledge as a network B_i of KUs $\{K_{ij}\}$, which are connected by edges $\{e_{i(j,k)}\}$ based on

---

[16] By way of example, the topology of the innovation network A was chosen to be a random Erdős–Rényi graph, and each agent was initialized with 10 KUs generated from a uniform distribution.

[17] Consequently, all curves shown in Fig. 5.3 would collapse onto a single master curve when applying a γ-dependent scaling factor on the time axis.

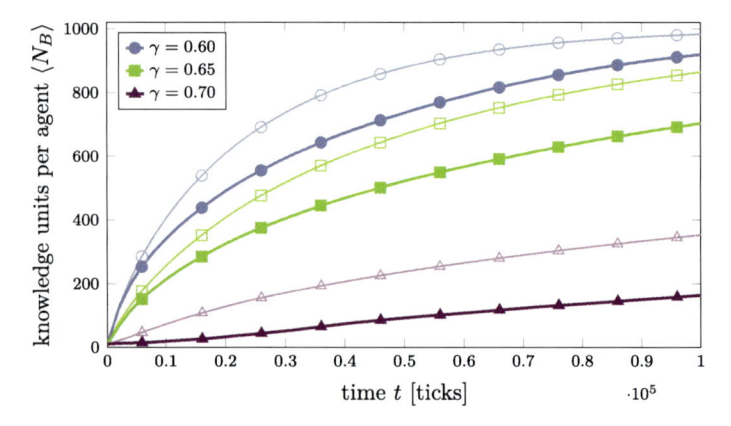

Fig. 5.4 Average number of KUs $\langle N_B \rangle$ per agent over time in an innovation network A with an Erdős–Rényi (random network) topology for different compatibility thresholds γ. Thick lines with filled symbols are obtained from simulations with the random walk within the agent's knowledge networks switched on, while thin lines with corresponding empty symbols and colors are taken from simulations with unweighted random jumps. Lines are averages over 100 agents. The standard error of the mean is smaller than the line width and therefore not shown.

their pairwise compatibility c_{jk} (Eq. (5.1)) under the γ-dependent condition P_{jk} (Eq. (5.3)). This allows us to switch on the weighted random walks of the agents' shifting attention or focus along these edges, with the consequence that KUs that are strongly connected to other related KUs have a higher probability to attract attention, and, therefore, also a higher probability of being transmitted.

As mentioned above, Fig. 5.4 shows the average number $\langle N_B \rangle$ of KUs per agent over time for different values of γ. Curves with thick lines and filled symbols are obtained from simulations with the weighted random walk within the agent's knowledge networks switched on (knowledge exploitation strategy), while thin lines with corresponding empty symbols and colors are taken from simulations with unweighted random jumps (knowledge exploration strategy).[18]

The first and obvious observation clearly is that the curves from simulations with knowledge exploitation (i.e., random walk switched on; thick lines with filled symbols) generally lie below the corresponding curves obtained from simulations where all agents follow the exploration strategy (i.e., random jumps; thin lines, empty symbols, same color code). This indicates that the exploitation strategy of agents focusing on well-connected KUs (i.e., those exhibiting strong compatibility relations with others within the agents' knowledge networks) causes a slowdown in the dynamics of knowledge diffusion between different agents. More precisely, this will cause neighboring agents to exchange knowledge only in a limited field, thereby

---

[18] Due to the fact that for low compatibility thresholds γ the resulting knowledge networks are extremely dense, the memory requirements of the simulations become prohibitively large. We therefore had to restrict our investigations to values of $\gamma \geq 0.6$.

successively decreasing their chance that they can learn something new, causing their attention to be "paradigmatically locked in" in a truly Kuhnian sense.[19]

Moreover, the average growth rates now exhibit a slightly different time dependence compared to the case with random jumps (exploration).[20] While at the very beginning of the simulation the curves for $\gamma = 0.60$ and $\gamma = 0.65$ (blue lines with squares and green lines with circles, respectively) almost coincide with the corresponding curves from simulations with random jumps, they quickly start to deviate. In contrast to those two cases, for $\gamma = 0.7$, the beginning of the curve with the random walk switched on (thick purple line, filled triangles) is convex. As we will see at a later point in this work, this specific behavior originates from the fact that some agents in the simulation may be unable to find any compatible knowledge among their neighbors in the early stages of the simulation if the compatibility threshold is rather high.

5.3.2 Effects of Knowledge Diversity

It has been observed that innovation networks, sectors, regions, or industries often exhibit different and uneven developments also in terms of knowledge, whereby it becomes increasingly important to better understand the factors contributing to these diverging knowledge trajectories (e.g., Feldman and Audretsch 1999; Foray 2004, 2014; Frenken et al. 2007; Smith 2000; Vermeulen and Pyka 2017). Therefore, in this section, we extend our baseline model to analyze if and how the (initial) diversity of knowledge bases in an innovation network can influence the knowledge diffusion process.

So far, we have only considered systems where KUs originate from the same uniform distribution for all agents in the innovation network. Even though, for this setup, the dependence of knowledge exchange on compatibility as well as the introduction of the random walk results in a statistically significant quantitative change of knowledge diffusion dynamics, the qualitative behavior is still very similar to the case where compatibility has no influence on knowledge exchange at all. As we will see in the following subsections, this will change drastically if we consider different levels of knowledge homo- and heterogeneity within and among agents.

In the following two paragraphs, we construct two different scenarios of knowledge diversity. In the first scenario, we imagine an example where all agents essentially share the same knowledge background, for example, because they are rooted within the same technological field. In other words, in this first scenario, all agents' knowledge networks share the same "ancestral" knowledge unit (AKU). In the sec-

---

[19] As, for example, Chris Buskes (2010) explains with reference to Thomas Kuhn's *The Structure of Scientific Revolutions* (1996), the more specialized the members of a particular scientific community, the more difficult it will be for the paradigms to "interbreed" (not unlike members of different biological species).

[20] Because of this, it is no longer possible to collapse the different lines onto a single master curve by means of time rescaling.

ond scenario, we consider a situation where agents do not share a common AKU, for example, because the innovation network consists of agents from different technological fields.

5.3.2.1 Common Knowledge Background

The representation of KUs as bit strings offers the possibility to control their average pairwise compatibility by employing different methods of bit string generation. If all KUs are generated randomly from a uniform distribution, the distribution of compatibilities between all pairs of such KUs has the shape of a Gaussian centered at $c = 0.5$. If the integral of the compatibility distribution is normalized to 1, the distribution can be interpreted as a probability density function $PDF(c)$ (see black line with filled circles in Fig. 5.5). Obviously, this distribution remains intact during the entire course of the simulation, since each newly acquired KU originates from the same initial uniform distribution.

In order to implement the scenario where agents have a common knowledge background, the average pairwise compatibility c_{mn} between pairs of KUs K_{im}, K_{jn} belonging to any two different agents a_i, a_j has to be shifted closer to one. This can be accomplished by initially supplying all agents in the system with the same AKU (arbitrarily chosen for each simulation run during initialization). Each KU of an agent is then derived from a copy of the AKU by reassigning a random value to each bit of the copied AKU with a reassignment probability p_k. Accordingly, for $p_k = 0$, no bits would be changed and, thus, all KUs would equal the AKU. In contrast, for $p_k = 1$, each KU would be drawn from a uniform random distribution again, which would lead to the same overall compatibility distribution as before. To illustrate this behavior, Fig. 5.5 depicts compatibility distributions resulting from different values of p_k.

Since the AKU is the same for each agent in A, these compatibility distributions represent compatibilities between all pairs of KUs within the whole simulated system, i.e., between and within agents. Consequently, a lower value of p_k implies a higher knowledge homogeneity in the system.

Figure 5.6 shows the diffusion performance, again measured by the average number of KUs per agent over time. As before, we compare the results for exploitation and exploration strategies (i.e., with and without our weighted random walk). The first observation is that, for higher knowledge homogeneity (systems with lower p_k), the learning process is much faster since it is easier for the agents to find compatible knowledge. The second observation is that, in contrast to the previous case (Fig. 5.4), for all depicted levels of knowledge heterogeneity, initially, a knowledge exploitation strategy (weighted random walk on) means that learning is faster but, after a certain point in time (earlier for lower p_k), knowledge exploration becomes more efficient. If an agent focuses on a highly compatible subset of its knowledge, the average compatibility with such a subset focused by another agent is comparatively high, since agents have a common knowledge background. KUs belonging to these subsets can therefore be transferred very rapidly at the beginning. Consequently, we see that the

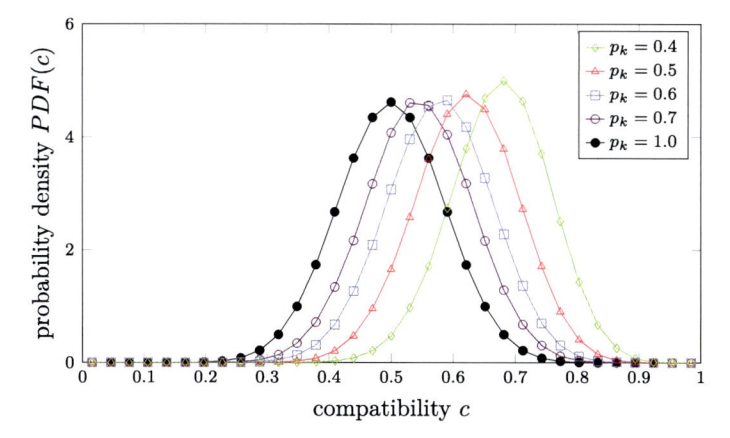

5 It's a Match! Simulating Compatibility-based Learning in a Network of Networks   115

Fig. 5.5 Distribution of compatibilities between pairs of KUs with $n_K = 32$ bits generated from an AKU with different bit reassignment probabilities p_k. The area under the curves is normalized to 1 so that the curves can be interpreted as probability density functions (PDFs). The case $p_k = 1$ (black line with filled circles) corresponds to the case where each knowledge unit is completely random and independent of the AKU.

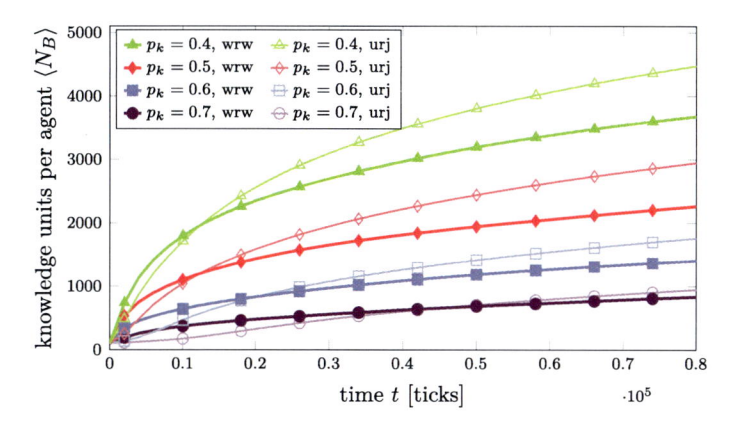

Fig. 5.6 Average number of KUs $\langle N_B \rangle$ per agent over time in an innovation network A with an Erdős–Rényi (random network) topology for different KU bit reassignment probabilities p_k. The compatibility threshold was set to $\gamma = 0.75$ in all simulations. Thick lines with filled symbols are obtained from simulations with the random walk within the agent's knowledge networks switched on (wrw), while thin lines with corresponding empty symbols and colors are taken from simulations with unweighted random jumps (urj). Lines are averages over 100 agents. The standard error of the mean is smaller than the line width and therefore not shown.

"paradigmatic lock-in" described above may actually be beneficial at the beginning of the diffusion process if agents share a common knowledge background. However, we can still observe that this lock-in caused by exploitation leads to a situation where knowledge diffusion slows down and thus, in the long run, exploration will be more conducive to knowledge diffusion.[21]

5.3.2.2 Different Knowledge Background

In contrast to the previous case, we now consider a scenario where we no longer assume a common knowledge background for all agents in the system, for example, because actors in the innovation network come from different technological fields. In the model, instead of deriving each knowledge network from *the same* AKU, agents derive their knowledge networks from *diverse* AKUs now.

We achieve this diversity in knowledge background by reassigning a random value to each bit of an agent's AKU with an AKU bit reassignment probability of p_a while keeping the KU bit reassignment probability constant at $p_k = 0.6$. This means that for $p_a = 0$, we would have the same scenario as before, and each agent would have the same AKU, whereas for $p_a = 1$, each agent has a random AKU. More precisely, in our example, p_k can be interpreted as a measure for the heterogeneity of knowledge within a particular technological field, whereas p_a determines the heterogeneity across technological fields within an innovation network. Since p_k is fixed to 0.6 in this scenario, increasing p_a essentially means increasing the average cognitive distance between agents.

In Fig. 5.7, we show the results for different degrees of average cognitive distance, ranging from $p_a = 0.4$ to $p_a = 1.0$. As above, the curves depict the (average) growth of agents' knowledge networks over time. The first and probably most obvious thing to observe is that learning is much slower and the agents' average number of KUs is considerably lower than in the previous scenario with the common knowledge background.[22] Yet, the most striking result compared to the previous case is that for higher p_a, i.e., for situations where the AKUs of agents differ considerably ($p_a = 0.8$ and $p_a = 1.0$), knowledge exploration is always "better" in terms of knowledge network growth than knowledge exploitation. However, for moderate levels ($p_a = 0.4$ and $p_a = 0.6$), we see that exploitation leads to a faster learning process in the short and medium terms. In other words, introducing different technological fields within a network—and thereby increasing the initial diversity of knowledge between agents—does not lead to a situation where an exploration strategy is always advantageous. Instead, we see that exploitation is more beneficial (at least in the short and medium terms) in networks with relatively compatible but still different technological fields.

---

[21] In the words of Buskes (1998, p. 125), "If the previously trusted corpus of knowledge is left behind, the resulting extended search space may soon provide a new bridgehead for further research in the unfamiliar epistemic setting."

[22] For example, where for $p_k = 0.6$ a knowledge exploitation (random walk) strategy yielded about 1,000 KUs after roughly 40,000 ticks in the previous scenario, reaching 1,000 KUs in the current scenario requires around five times as long with the same strategy, even if diversity of knowledge backgrounds is relatively low ($p_a = 0.4$).

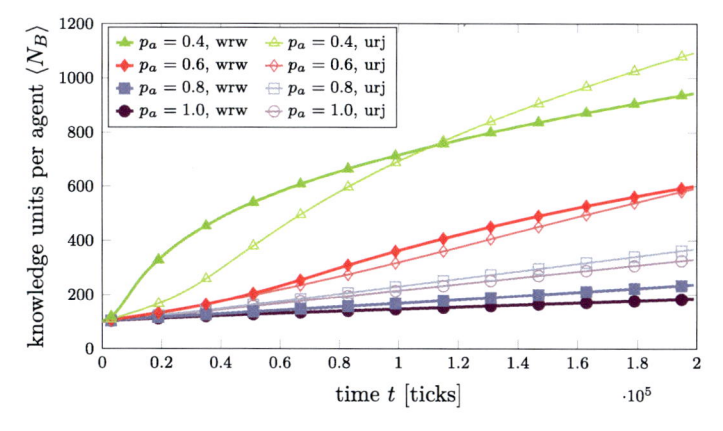

Fig. 5.7 Average number of KUs $\langle N_B \rangle$ per agent over time in an innovation network A with an Erdős–Rényi (random network) topology for different AKU bit reassignment probabilities p_a. The KU bit reassignment probability was set to $p_k = 0.6$ and the compatibility threshold to $\gamma = 0.75$ in all simulations. Thick lines with filled symbols are obtained from simulations with the random walk within the agent's knowledge networks switched on (wrw), while thin lines in corresponding colors with empty symbols are taken from simulations with unweighted random jumps (urj). Lines are averages over 1,000 agents. The standard error of the mean is smaller than the line width and therefore not shown.

5.3.3 Extended Analysis: Approaching the Qualitative Dimension of Knowledge

Up to this point, we have restricted ourselves to assessing overall quantitative changes in terms of the average number of KUs per agent, which gives us a measure for successful knowledge diffusion and, especially, assimilation processes. However, by changing the way knowledge representation is modeled from numbers and vectors to networks, we have arrived at a point where we can also re-conceptualize how the *effects* of knowledge diffusion are analyzed. As our model allows us to not just deal with agents' higher or lower knowledge "stocks" or knowledge "levels," we can shift the focus from cardinal measures referring to "more" knowledge, to an extended, qualitative, i.e., structural, analysis at the knowledge network level (B_i). A large number of different approaches and indices exist to describe a network's structural properties (e.g., Barabási 2016; Newman 2010; Wasserman and Faust, 1994). However, although well established for the description of social and other complex networks, an application to knowledge networks is not straightforward, particularly because the knowledge networks in our simulation are dynamic and may contain disconnected components, whereas many measures can only be meaningfully applied to static networks and connected graphs. Moreover, some measures such as the average path length are size-dependent and rather meaningless for growing knowledge networks with disconnected components, whereas other measures such as clustering coefficients are hard to interpret.[23] Nevertheless, there exist some network characteristics that can be used to illustrate the merits of our model.

In the following analysis of the structure of knowledge networks, we focus on the number of KUs, average degree, (weighted) density, and modularity of the knowledge networks. As before, the number of KUs is an important measure for knowledge diffusion performance. The average degree of KUs is, of course, also size-dependent and indicates the number of KUs in that network with which an average KU is compatible. The weighted density of a knowledge network indicates how well an agent can, on average, put its KUs into context. Modularity is a measure used for community detection in networks (e.g., Francisco and Oliveira 2011; Newman 2004a, 2004b; Newman and Girvan 2004; Sobolevsky et al. 2014). In the case of our model's knowledge networks, modularity may be a reasonable way to capture an agent's relative degree of knowledge *diversification* (high modularity) or *specialization* (low modularity in conjunction with high weighted density).[24]

Figure 5.8 shows the results for these indicators as an average over 10 simulation runs (i.e., 1,000 agents) for different degrees of initial cognitive distance (varying p_a). In this scenario, all agents pursue knowledge exploitation (weighted random walk on). As we have already seen above (in Fig. 5.7), learning is slower for higher p_a. Consequently, the average degree and the weighted density also increase much faster in the case of lower average cognitive distance, as it is harder for more diverse agents to "make sense" of the knowledge they receive from agents with a different knowledge background. Although all agents engage in knowledge exploitation (weighted random walk), by looking at the modularity we can see that in cases where average cognitive distance is lower (smaller p_a), modularity decreases, pointing to a tendency to specialize around a particular field, whereas for higher average cognitive distances (higher p_a), modularity increases and points to a more "clustered" or diversified knowledge base.[25]

---

[23]Even though various measures such as centralities may be applied to describe the "importance" of particular KUs in the knowledge networks by means of their centrality distributions, discussions of meaningful normalization and interpretation would go beyond the scope of the article and should be done in future work.

[24]Note that there may be a hypothetical case where two knowledge networks have the same low modularity, one of which has very low edge weights (i.e., consists of rather incompatible KUs) and the other has high edge weights (highly compatible KUs). In this hypothetical case, modularity cannot be meaningfully interpreted as a measure of diversification or specialization taken by itself. However, due to the compatibility threshold γ in our model, the extreme cases with low edge weights do not exist in our simulation.

[25]One may wonder whether these results may be a mere size effect. However, if we compare the structural properties of knowledge networks of the same size originating from different setups of cognitive distance, we clearly see that this is not the case. For example, for $p_a = 0.6$ and $p_a = 0.8$, a knowledge network size of a little more than 200 is achieved around time 70,000 and 200,000, respectively. Looking at the corresponding modularities gives us for $p_a = 0.6$ at $t = 70,000$ a modularity Q of about 0.6 and for $p_a = 0.8$ at $t = 200,000$ a Q of about 0.7.

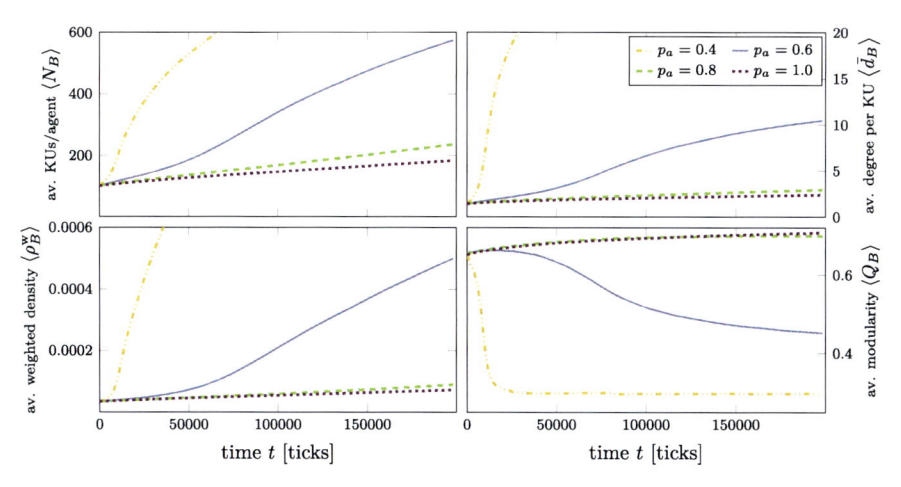

5 It's a Match! Simulating Compatibility-based Learning in a Network of Networks 119

Fig. 5.8 Change of knowledge network properties over time in an innovation network A with an Erdős–Rényi (random network) topology for different AKU bit reassignment probabilities p_a. The lines are averages over 1,000 agents. The KU bit reassignment probability was set to $p_k = 0.6$ and the compatibility threshold to $\gamma = 0.75$ in all simulations. Knowledge exploitation by all agents (wrw). Left: Average number $\langle N_B \rangle$ of KUs in $\{B_i\}$ and average weighted network density $\langle \rho_B^w \rangle$ of $\{B_i\}$. Right: Average degree $\langle \bar{d}_B \rangle$ of KUs in $\{B_i\}$ and average modularity $\langle Q_B \rangle$. The standard error of the mean is $\pm 1.5 \cdot 10^{-5}$ for $\langle \rho_B^w \rangle$ and in the order of the line width for the other three properties.

5.3.4 The Importance of Being in the Right Place at the Right Time

Although researchers and policy-makers may often be interested in the overall "performance" of a particular region or system (see, e.g., Foray 2014; Vermeulen and Pyka 2017), it should be kept in mind that the simulation results shown above (e.g., Fig. 5.8) are averaged over 10 simulation runs (i.e., 1,000 agents). However, by averaging time series of dynamic quantities, many details of the underlying dynamics might be obscured. In order to illuminate more details of individual knowledge network dynamics, we, therefore, show the individual time series for the number of KUs of each agent in one simulation in Fig. 5.9.

As we can see, agents' learning trajectories differ. In fact, at the beginning of the simulation, there are agents who are unable to find any compatible knowledge. This results in the convex shape of some of the averaged curves in Figs. 5.6, 5.7, and 5.8. To investigate the origins of these individual learning differences, we looked at the dependence of the individual agents' knowledge networks on their local degree, clustering coefficient, betweenness centrality, and harmonic closeness centrality. We found that a higher degree in A has a positive impact but with decreasing returns to scale. Harmonic closeness centrality was found to have the strongest impact, followed by degree and betweenness centrality. In contrast, an agent's local clustering

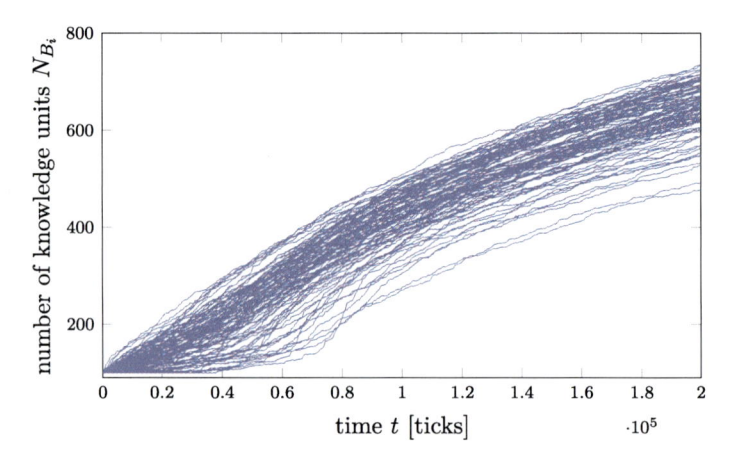

Fig. 5.9 Change of individual knowledge network sizes N_{B_i} over time in an innovation network A with an Erdős–Rényi (random network) topology for a moderate level of cognitive distance ($p_a = 0.6$ and $p_k = 0.6$), knowledge exploitation by all agents and $\gamma = 0.75$.

coefficient was found to have only a small but negative impact on an agent's learning performance.[26]

As an example, Fig. 5.10 shows the time series of relative deviations from the mean of different properties in B_i depending on agents' harmonic closeness centrality H in A in a scenario where all agents have a common knowledge background ($p_a = 0$) and engage in knowledge exploitation (weighted random walk). For the analysis, we first categorize agents into two groups, one containing agents with H above the median and the other containing agents with H below the median. Second, we compute the average relative deviations for each group from the mean of the entire population for the different properties of B_i mentioned before (N_B, $\bar{d}_B$, ρ_B^w, and Q_B) according to

$$\Delta X = \frac{1}{g} \sum_{i=1}^{g} \frac{X_i - \langle X \rangle}{\langle X \rangle}, \tag{5.4}$$

where $X \in \{N_B, \bar{d}_B, \rho_B^w, Q_B\}$, g is the number of agents belonging to one of the groups (below or above median H), and the operator $\langle \cdot \rangle$ denotes the average over all agents in the simulation, regardless of their group membership. Note that, according to Eq. (5.4), the different ΔX reported in the subgraphs of Fig. 5.10 are *group averages* of the relative deviation from the mean of the entire population.

In Fig. 5.10, the large positive and negative peaks of $\Delta \bar{d}_B$, $\Delta \rho_B^w$, and ΔQ_B at the beginning of the simulation are due to a combination of two things: First, agents with a high value of H are, on average, more likely to also have a higher degree. This means that they have a higher probability of finding compatible knowledge among their neighbors and can learn more rapidly. Second, knowledge networks are

---

[26] See Appendix C for statistical tests confirming these findings.

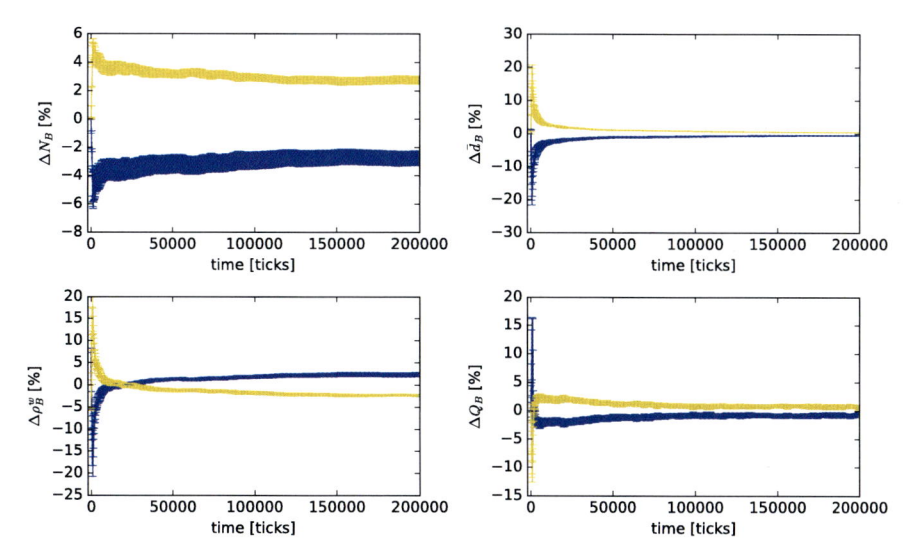

5 It's a Match! Simulating Compatibility-based Learning in a Network of Networks   121

Fig. 5.10 Time series of relative deviations from the mean of different knowledge network properties depending on agents' harmonic closeness centrality H in the innovation network A for the case of a common knowledge background ($p_a = 0.0$) and a moderate level of knowledge heterogeneity ($p_k = 0.6$). Investigated properties of the agents' knowledge networks are size N_B (top left), average degree $\bar{d}_B$ per KU (top right), weighted density ρ_B^w (bottom left), and modularity Q_B (bottom right). Yellow lines show averages for agents belonging to the group with H above the median and blue lines for agents belonging to the group with a value of H lying below the median. The data were computed from a system of 100 agents. Knowledge exploitation by all agents and $\gamma = 0.75$. Error bars represent the standard error of the mean of each group.

initially very small (100 KUs), with the effect that an additional KU can significantly reduce the modularity of the knowledge network. Consequently, since we are dealing with *relative* deviations from the mean, the curves of the other group (with low H) show the opposite behavior. In summary, the data presented in Fig. 5.10 imply that if the agents in an innovation network with moderate levels of knowledge heterogeneity ($p_k = 0.6$) have a common knowledge background ($p_a = 0$), in the long run, short communication paths to the other agents in the network (higher H) result in a small but almost constant advantage in learning with slightly higher knowledge diversification.

Figure 5.11 also shows the time series of average relative deviations from the mean of different properties in B_i depending on H in A; however, this time for a scenario with highly diverse knowledge backgrounds ($p_a = 0.8$; all other parameters are equal to the previous case presented in Fig. 5.10). As we can see, in the long run, in innovation networks with agents from initially highly different knowledge backgrounds, "closer" agents (with H above the median) have, on average, a significant advantage in the number of KUs and their KUs are also more compatible to others in their knowledge network (higher $\bar{d}_B$). In other words, in innovation networks with actors from diverse knowledge backgrounds, agents with fewer and longer com-

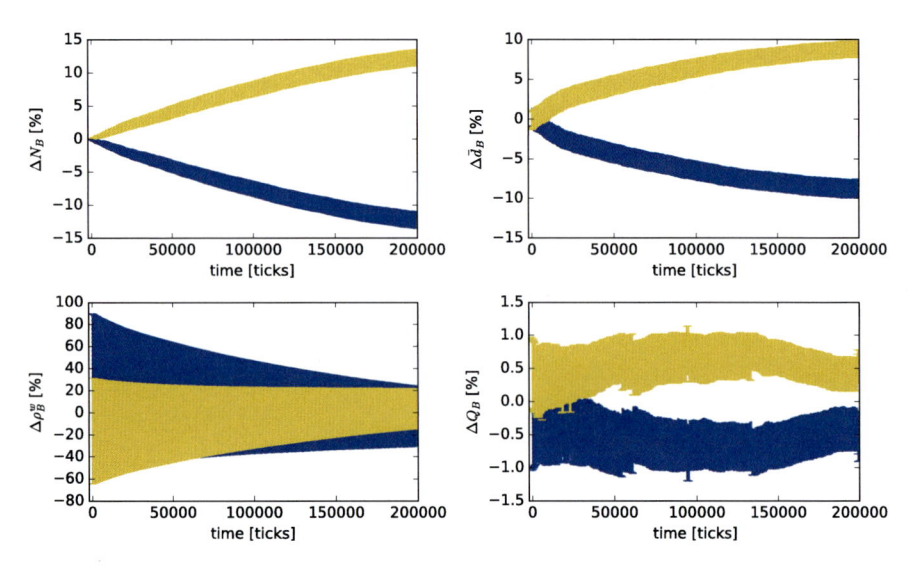

Fig. 5.11 Time series of relative deviations from the mean of different knowledge network properties depending on agents' harmonic closeness centrality H in the innovation network A for the case of highly diverse knowledge backgrounds ($p_a = 0.8$) and a moderate level of knowledge heterogeneity ($p_k = 0.6$). Investigated properties of the agents' knowledge network are size N_B (top left), average degree $\bar{d}_B$ per KU (top right), weighted density ρ_B^w (bottom left), and modularity Q_B (bottom right). Yellow lines show averages for agents belonging to the group with H above the median and blue lines for agents belonging to the group with a value of H lying below the median. The data were computed from a system of 100 agents. Knowledge exploitation by all agents and $\gamma = 0.75$. Error bars represent the standard error of the mean of each group.

munication paths have a higher chance that incompatible (i.e., cognitively distant) agents effectively block their transmission of compatible knowledge. In contrast, learning processes of rather well-connected agents with high H are less likely to be "stalled" by other agents. Finally, in Fig. 5.11, modularity does not show significant differences, with the implication that—although advantageous for learning—shorter communication paths do not seem to influence an agent's propensity for knowledge diversification or specialization.

5.3.5 Results for Different Innovation Network Properties

As we have found that the agents' local properties or positions in the innovation network A can influence their learning processes, we also expect the global properties of A to have an impact on overall diffusion performance. This is in line with many models of knowledge diffusion on networks that have, for various reasons, focused on analyzing the speed and extent of the diffusion process depending on different topologies of the agents' social network (e.g., Bogner et al. 2018; Buchmann and

Pyka 2012; Cowan and Jonard 2004; Morone et al. 2007; Mueller et al. 2014, 2017; among others). In this section, we, therefore, analyze different (static) structures of A and compare the dynamics at the knowledge network level (B_i) depending on different types or properties of A.[27]

Up to this point, innovation networks in our analyses were assumed to be an **Erdős–Rényi** (ER) random graph (Erdős and Rényi 1959, 1960). Here, we additionally use the other two most widely used network types of **Barabási-Albert** (BA) (Barabási and Albert 1999, 2002) and **Watts–Strogatz** (WS) (Watts and Strogatz 1998) to compare the resulting dynamics in B_i for different structures of A.

As already mentioned above, for the sake of comparability, we fix the number of agents and the number of links between them for all network types. We additionally require the networks to not have any disconnected components.[28] Table 5.1 shows how the networks differ in their properties. Judging from our previous results, we expect networks with a high average harmonic closeness centrality H and a low clustering coefficient to perform best in terms of knowledge network growth. Additionally, since an agent's local degree showed "decreasing returns to scale," networks with a high median degree should be advantageous for knowledge diffusion. If we compare the properties of the different types of A listed in Table 5.1, we expect the following: both BA and ER have a relatively high H and relatively low clustering coefficients as well as comparatively small network diameters and should, therefore, facilitate knowledge diffusion more efficiently than the WS network, so that the expected "ranking" is ER ≈ BA > WS. However, due to our focus on compatibility-based learning and the simple knowledge trade protocol we employ, we do not expect to see fundamental qualitative differences in the average diffusion performance. Nevertheless, a closer look at the dynamics of the properties of B_i may reveal significant quantitative differences.

In Fig. 5.12, the left panel (a) shows that in a system where agents share a common knowledge background ($p_a = 0$), ER innovation networks perform slightly better than BA and WS. The right panel (b) in turn shows that in a system with highly diverse knowledge backgrounds ($p_a = 1$), BA networks are more efficient than ER

---

[27] We are aware that, in reality, innovation networks are often dynamic because alliances between agents may be discontinued due to increasing cognitive overlaps and (re-)established when partners exhibit a sufficient cognitive distance (see also Egbetokun and Savin 2014, their footnote 7, and the discussion in Cowan et al. 2006). However, the static structure of the innovation networks is necessary at this stage to separate structural effects at the level of A from structural effects at the knowledge network level (B_i). Furthermore, static networks A can be interpreted as formal networks that are stable over time due to geographically and contractually binding factors (Bogner et al. 2018; Buchmann and Pyka 2012).

[28] Especially in the case of random ER networks, simply generating a random network of N_A agents and M_A edges poses problems, since the resulting network might contain disconnected components. In that case, one would not simulate a single innovation network of size N_A, but several separate networks with sizes smaller than N_A. In order to avoid this problem, we require the networks to consist of one single component. These requirements impose restrictions on the respective generating algorithms, which are discussed in Appendix D. For WS, we used a rewiring probability of $p_{WS} = 0.1$ for the network to exhibit small-world properties (Watts and Strogatz 1998).

Table 5.1 Several global properties of the examined innovation networks A. Values are averaged over 100 simulations with networks set up as described in Appendix D. Uncertainties of the last digits are given in parentheses and represent the standard error of the mean.

	Barabási–Albert	Erdős–Rényi	Watts–Strogatz
Number of agents N_A	100	100	100
Number of links between agents M_A	200	200	200
Network density ρ_A	0.04	0.04	0.04
Average degree D	4.0	4.0	4.0
Network diameter $\varnothing$	5.34(5)	7.36(7)	10.6(1)
Average path length l	2.952(8)	3.455(5)	5.10(4)
Average clustering coefficient C_l	0.160(4)	0.037(1)	0.377(3)
Global clustering coefficient C_g	2.32(3)	0.60(2)	4.45(3)
Average betweenness centrality C_B	0.01992(8)	0.02506(5)	0.04182(5)
Average harmonic closeness centrality H	37.39(8)	33.05(3)	25.0(1)
Degree distribution	Power law	Poisson	Narrow Poisson-like

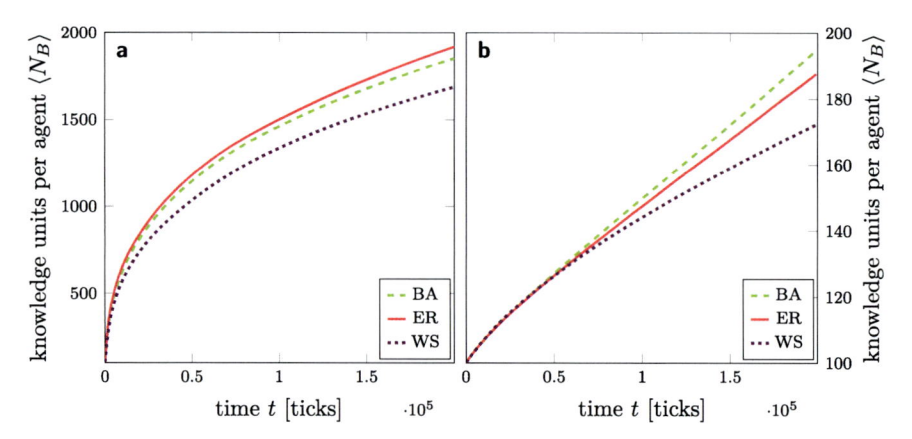

Fig. 5.12 Time series of average knowledge network sizes $\langle N_B \rangle$ for different innovation network topologies. The different topologies are Barabási–Albert (BA), Erdős–Rényi (ER), and Watts–Strogatz (WS). The lines are averages over 50 simulations with 100 agents (5,000 agents in total) per topology. **a**: System of 100 agents with a common knowledge background ($p_a = 0$) and a moderate level of knowledge heterogeneity ($p_k = 0.6$). **b**: System of 100 agents with completely different knowledge background ($p_a = 1$) and a moderate level of knowledge heterogeneity ($p_k = 0.6$). All agents follow a knowledge exploitation strategy; the compatibility threshold is $\gamma = 0.75$. The standard error of the mean is smaller than the line width and therefore not shown.

and WS. The switch between the diffusion performance of ER and BA in scenarios a and b also relates to the discussion in the literature about the ambiguous effects of "stars" (i.e., agents with a high degree) on knowledge diffusion, depending on them freely giving away their knowledge or trading it (e.g., Bogner et al. 2018; Cowan and Jonard 2007; Müller 2017, Chap. 5; Mueller et al. 2017). Our results are, therefore, also relevant for the discussions about the importance of heterogeneity and diversity of agents within the networks. Hence, with regard to the different performance of the BA network, we can argue that the effect of stars does not just depend on whether they are givers or traders of knowledge but also on the knowledge diversity in the system.

However, comparing these rather artificial network structures may obscure another important element that has received relatively little attention in the literature on innovation networks, namely, that alliances between agents are often formed on the basis of complementary knowledge between partners (e.g., Baum et al. 2010; Cowan and Jonard 2009; Tur and Azagra-Caro 2018). Therefore, in line with the ideas of Joel Baum, Robin Cowan, and Nicolas Jonard (2010), we now additionally compare the diffusion performance in a "compatibility-based" (CB) innovation network that is created with the same number of edges as the other innovation networks based on the assumption that agents must have a certain fit or compatibility in their knowledge to form an alliance (see also Cowan and Jonard, 2009).[29] At this point, we assume that the alliances thus formed remain stable during the simulation for the sake of comparability and, therefore, the CB network also remains static. Yet, we are aware of the fact that cognitive distances between agents may change over time.

As we can see in Fig. 5.13, in a scenario where agents have a very different knowledge background, learning in the CB network is much more efficient—and expectedly so—than in the other three network types, as it is much easier for agents to find compatible knowledge among their neighbors. This result is also highly relevant for the overall discussion on the role of efficient network structures, complementary knowledge, and social capital (e.g., as discussed by Baum et al. 2010; see also Cowan et al. 2006, on a related note). Put differently, it may be futile to search for an optimal network structure if the measure of optimal performance exhibits a strong dependence on the properties of individual agents in the network. This finding is not only relevant for our model, where compatibility between agents is implicitly determined by the average compatibility of their knowledge networks. In fact, even in models where knowledge compatibility is not explicitly incorporated, it might still affect the dynamics implicitly. For example, even a simple barter trade mechanism implies some sort of compatibility, since agents have to somehow mutually agree on exchanging knowledge. This, in turn, must be determined by some criterion which is likely to exhibit effects similar to the explicitly incorporated notion of compatibility in our model.

---

[29] For details on how these CB networks are created, see Appendix D.

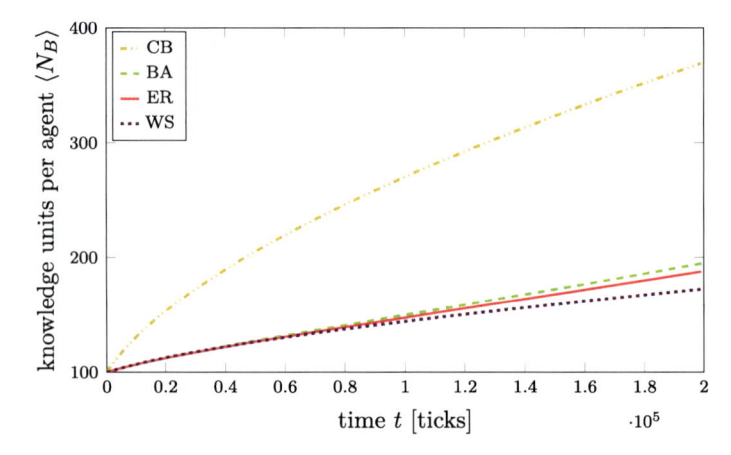

Fig. 5.13 Time series of average knowledge network sizes $\langle N_B \rangle$ for different innovation network topologies. The different topologies are Barabási–Albert (BA), Erdős–Rényi (ER), Watts–Strogatz (WS), and compatibility-based (CB). The lines are averages over 50 simulations with 100 agents (5,000 agents in total) per topology. Agents are initialized with a highly diverse knowledge background ($p_a = 1$) and a moderate level of knowledge heterogeneity ($p_k = 0.6$). All agents follow a knowledge exploitation strategy; the compatibility threshold is $\gamma = 0.75$. The standard error of the mean is smaller than the line width and therefore not shown.

5.4 Conclusion and Outlook

With this article, we contribute to the literature on knowledge dynamics in innovation networks. Since knowledge and learning trajectories differ between and within industries, sectors, and regions, models are needed that aid researchers and policy-makers in comprehending these complex processes and their dependence on different initial conditions. Although knowledge is and always has been a somewhat elusive concept, recent advances in network science and (computational) economic modeling have already contributed a lot to better understand diffusion phenomena. Nevertheless, as elaborated in the introductory and theoretical sections of this article, several important characteristics of knowledge and their implications for modeling diffusion have not been explicitly discussed so far. We have addressed this research gap by developing an agent-based simulation model that captures knowledge diffusion in an innovation network and assimilation depending on compatibility and a shifting focus of attention among networked units of knowledge. This more fine-grained model of compatibility-based, path-dependent learning processes in innovation networks allows us to get a little closer to the qualitative dimension of knowledge representations than some of the other models developed in this field.

So, what have we gained from this added complexity? First, we have been able to analyze the effects of knowledge diversity in different scenarios against the backdrop of agents' knowledge exploitation versus exploration strategies. Although models are always just a simplification of reality, and there are many other factors to be considered in a real innovation system,[30] the results of our simulation can already provide us with some relevant insights by pointing toward previously disregarded effects. More precisely, our results relate, among others, to discussions about (cognitive) proximity (e.g., Boschma 2005; Nooteboom et al. 2007) and support the claims that the diffusion of knowledge in innovation networks strongly depends on the diversity of knowledge available in the system. With this model, we are able to distinguish between different levels of knowledge diversity. Most notably, diversity of knowledge *between* agents is also present in previous models of knowledge diffusion; yet, to the best of our knowledge, in previous models, the heterogeneity of KUs *within* agents does not influence which part of their knowledge base is more likely to be selected for transmission and, in turn, assimilated. As a consequence of this added layer of complexity, we are now able to see that the advantageousness of knowledge exploitation versus exploration strategies in terms of knowledge diffusion performance differs considerably depending on the *level* of knowledge diversity. We have observed that in a scenario where agents share a common knowledge background (e.g., because they are rooted within the same technological field), an exploitation strategy is always better in the beginning, whereas exploration leads to more efficient knowledge diffusion in the medium and long run. When agents do not share a common knowledge background, however, the effects are less clear-cut. For situations where the technological fields present within an innovation network exhibit moderate knowledge heterogeneities and where the individual knowledge backgrounds (e.g., technological fields) of the actors differ considerably, exploration is always conducive to their knowledge network growth, whereas for less diverse knowledge backgrounds, knowledge exploitation is more efficient, at least in the short and medium terms. Additionally, with our model, we have been able to measure knowledge diffusion differently by focusing on structural effects in knowledge networks (e.g., captured by average degree, weighted density, and modularity of knowledge networks). This is another step beyond the merely cardinal, cumulative measures of many previous models in terms of higher or lower knowledge levels in given categories.

Finally, after comparing knowledge diffusion performance depending on different innovation network topologies, we can see that our results have implications for both researchers and policy-makers interested in the knowledge diffusion performance of an innovation network. If the aim is to find "efficient" network structures to improve collective learning, it may—in some scenarios (e.g., in the case of knowledge exploitation by all agents)—be more important to connect agents with more compatible knowledge (i.e., lower cognitive distance) instead of just focusing on the structural (network) characteristics of the population of agents. Consequently, discussions about an "optimal" network structure should likewise revolve around cognitive distance and knowledge diversity, which ideally also includes diverse cultural knowledge (or a cultural background) that may be more or less compatible to particular areas of economically relevant knowledge. Based on the results of our simulation, we can, therefore, conclude that knowledge diffusion performance is not

---

[30] For example, it should be considered whether knowledge can be utilized for new product development, the (search) costs, whether knowledge is tacit or codified, etc.

only affected by the social structure (or network topology) of an innovation system, but also depends on the distribution of agents with respect to their individual properties (on that topology).

However, we also have to keep in mind that, for our analysis of innovation networks that are formed on the basis of a knowledge fit between agents (i.e., a compatibility-based network topology), we have only considered static innovation networks and different degrees of *initial* cognitive distance. Although we have been able to show that the diffusion performance differs considerably compared to other network topologies, the cognitive distances between agents change over time and, with this, the "optimal" innovation network structure should also change. Consequently, we could argue that if agents' properties are not static, neither should innovation networks be treated as such. In other words, an "optimal" network structure is very unlikely to be static. In this regard, our model can serve as a suitable starting point to investigate these issues in more detail in future research endeavors.

Although this model already yields pertinent insights, further work is not only possible but necessary to get closer to a complete picture of knowledge dynamics in innovation networks. Future research opportunities include additional analyses with the present model and extensions of the model itself. For example, one potentially insightful additional analysis with the present model could be "heatmaps" for edges in the innovation network in order to capture which communication paths have been used the most and why. This could be particularly interesting in the case where two connected agents have rather incompatible KUs so that we would expect to observe bottlenecks in knowledge diffusion regardless of their formal link.

Potentially informative model extensions include the following examples:

- The next plausible step would be to extend our model to study cases where the topology of the underlying innovation network is not static, i.e., actors can freely choose with whom they form and dissolve alliances. Consequently, in this case, innovation networks would be dynamic and co-evolving with knowledge diffusion and assimilation dynamics at the level of the agents' knowledge networks. In this context, it would also be interesting to fathom the most promising "strategy mixes" in terms of agents' knowledge exploitation to exploration ratios.
- Due to the fact that in our model compatibility between KUs is evaluated only between the receiving agents' currently focused KU with the potentially received one, it would also be attractive to compare the results once we implement the condition that the received KU has to be compatible also with adjacent KUs. In this way, we could capture diminishing marginal benefits of receiving additional KUs.
- Our model should also be extended to allow for differential retention and "forgetting" of knowledge (e.g., deletion of nodes in the knowledge network), for example, in order to analyze the effects of not using certain KUs. Additionally, in future extensions, KUs could also be deleted in favor of "updated" knowledge.
- Another interesting model extension (especially from the standpoint of memetics) may incorporate the creation of new KUs based on variation or recombination of old ones and other "breeding" mechanisms. In this regard, a promising way may

include merging our diffusion model with the mechanisms employed in graph-based knowledge creation models (e.g., Morone and Taylor 2010; Vermeulen and Pyka 2017).
- Additionally, the model could be upgraded to allow for and analyze the dynamics of production and consumption of new products (based on the knowledge networks of the agents) in line with previous models (e.g., Mueller et al. 2015; Schlaile et al. 2018).
- As our model uses a simple pull mechanism for knowledge exchange, it can also be extended to analyze more complex knowledge trade mechanisms (e.g., a barter trade or some kind of payment in return for knowledge) and to compare the results between these mechanisms.
- Innovation networks entail not only knowledge diffusion but also financial flows between agents (e.g., Buchmann and Pyka 2012). Future research should thus be aimed at improving our model by incorporating capital stocks and financial flows as well.
- Another extension that is somewhat related to the previous ones could incorporate competition between agents (e.g., firms) within the innovation network.[31]

Appendix

A) Choice of Model Parameters

The size of the innovation network A (i.e., the "macroscopic" network connecting different agents) is chosen as $N_A = 100$ in all simulations. This size is small enough to ensure computational feasibility, yet large enough to exhibit the characteristic features of the different network types employed in Sect. 5.3.5. However, in order to isolate the effect of network topology on diffusion dynamics to the greatest possible extent, we also fix the number $M_A = 200$ of undirected edges connecting the agents. Analogously, the "microscopic" intra-agent knowledge networks B_i containing an individual agent's KUs are initially set up with a size of $N_B = 100$ KUs per agent unless stated otherwise.

Our model employs bit strings as a representation of KUs and uses Eq. (5.1) as a measure for compatibility between any two of such units. However, this definition of compatibility is based on the Hamming distance, which poses problems when it comes to the generation of random KUs. If all KUs are generated randomly from a uniform distribution, the distribution of compatibilities between all pairs of such KUs has the shape of a Gaussian centered at $c = 0.5$. If the integral of the compatibility distribution is normalized to 1, the distribution can be interpreted as a probability density function $PDF(c)$. This, in turn, means that the probability p_c

---

[31] This extension was also suggested by Lorenzo Zirulia (2012) with regard to the model developed by Morone and Taylor (2010).

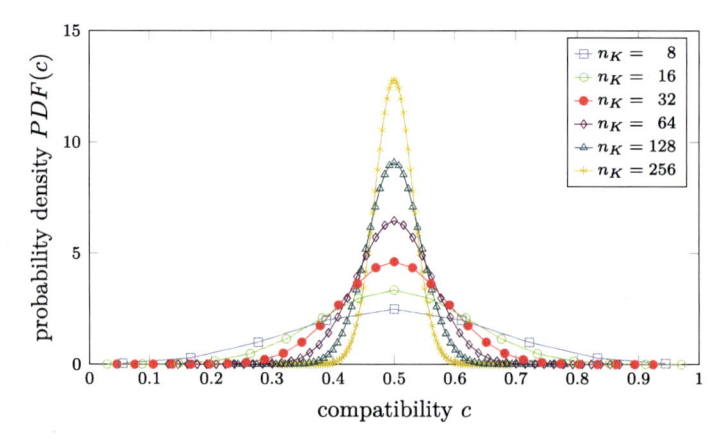

Fig. 5.14 Distribution of compatibilities between pairs of randomly generated KUs for different numbers of bits n_K per KU. The area under the curves is normalized to 1 so that the curves can be interpreted as probability density functions (PDFs). This, in turn, means that the probability p_c for the compatibility of two randomly generated KUs to lie between the threshold γ and 1 is $p_c(\gamma) = \int_\gamma^1 PDF(c)dc$. Note that for increasing n_K the resolution of the respective *PDF* increases (depicted by the points on each line), but its shape becomes narrower. Thus, the number n_K should be chosen with care.

for the compatibility of two randomly generated KUs to lie above the threshold γ is given as

$$p_c(\gamma) = \int_0^1 P(\gamma) PDF(c)\, dc \stackrel{(5.3)}{=} \int_\gamma^1 PDF(c)\, dc. \quad (5.5)$$

The width of the *PDF*, however, depends on the number of bits n_K per KU (see Fig. 5.14). For small values of n_K, the *PDF* is relatively broad. Since bit strings are discrete quantities, the number of different values the compatibility can take on is limited to $n_K + 1$, though. Thus, the discretization of the compatibility range is rather coarse for small n_K. Accordingly, the resolution of compatibility improves with increasing n_K. However, simply increasing n_K to a very large number is not advisable, since the Gaussian shape of the *PDF* narrows with increasing n_K, leaving only a very small range around $c = 0.5$ that actually contributes to the integral. In fact, the *PDF* approaches a Dirac delta function for $n_K \to \infty$. This behavior is illustrated in Fig. 5.14.

A comparison of simulations with different values of n_K showed that resolution is only an issue for very small n_K. Additionally, care must be taken that the maximum number of possible unique KUs $m_{\max} = 2^{n_K}$ is much greater than the number of unique KUs in the simulation, which is $m_{\text{sim}} \leq N_A \cdot N_B$. We found that $n_K = 32$ fulfills these requirements and a further increase of n_K leads to no qualitative change in the simulation results when the compatibility threshold γ is adjusted accordingly (i.e., it yields the same value of p_c).

Since we know the initial *PDF* for compatibilities in the system, we can explicitly calculate the probability of existence p_c for such edges for any threshold γ by evaluating Eq. (5.5). It thus becomes evident that the initial setup of the individual KU networks B_i corresponds to a *binomial random graph* (first introduced by Gilbert, 1959) constructed according to the $\mathbb{G}_{n,p}$-algorithm (see Frieze and Karoński, 2016) with $n = N_B$ and $p = p_c$, which is equivalent to an Erdős–Rényi random graph. The probabilities p_c thus correspond to the expected values of the (unweighted) initial network density $\langle \rho_B \rangle$.

B) Technical Framework

Our simulation is written mainly in the C++ programming language[32] and builds on the software infrastructure provided by the ABM framework repast HPC version 2.1 (Collier and North 2013), yielding a parallelizable and scalable code suitable for large-scale distributed computing platforms. The intra-agent dynamics are modeled using the Boost Graph Library (version 1.60.0) (Siek et al. 2002), and parallelization heavily relies on the Boost MPI and Boost Serialize libraries. All analyses of network properties were performed with self-written tools employing algorithms provided by the Boost Graph Library and a customized version of the Combo code (Sobolevsky et al. 2014). For the visualization of the simulated networks, the software Gephi (Bastian et al. 2009) is used. Randomness and reproducibility are guaranteed by giving each agent its own instance of a pseudorandom number generator (PRNG). The PRNG used here employs the *Mersenne Twister MT19937* algorithm (Matsumoto and Nishimura 1998) provided by the GNU Scientific Library (GSL version 1.16) (Galassi et al. 2009). Reproducibility is guaranteed by choosing a random *master seed* at the beginning of the simulation, from which the seeds for all the agents' private PRNGs are calculated in a deterministic way. The master seed is saved to disk together with all other relevant simulation parameters prior to the start of the simulation so that it is possible to re-run the very same simulation later.

C) Dependence of Knowledge Network Size on Local Properties in the Innovation Network

In order to quantify the dependence of the number of knowledge units N_B on the agents' location in the innovation network A, we calculated Pearson correlation coefficients of N_B with respect to the degree, local clustering coefficient, betweenness centrality, and harmonic closeness centrality of the corresponding agents in A. The Pearson correlation coefficient R assumes a linear relationship between the correlated

---

[32]The C++11 standard was used. Smaller parts of the simulation are written in C (C99 standard) and Assembly, and a part of the analysis tools are written in Python (Python 2.7 standard).

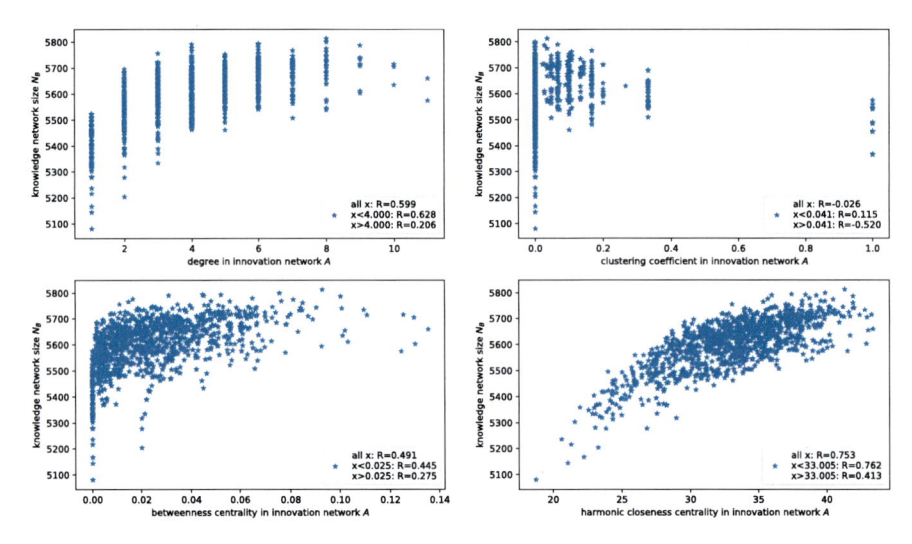

Fig. 5.15 Agents' number of KUs N_B with respect to their local properties in an innovation network A with Erdős–Rényi (random network) topology. The investigated local properties are degree (top left), local clustering coefficient (top right), betweenness centrality (bottom left), and harmonic closeness centrality (bottom right). The numbers R reported in the plot legends are Pearson correlation coefficients within the specified range. System parameters are $\gamma = 0.75$, $p_a = 0$, $p_k = 0.6$, knowledge exploitation strategy (random walk on), 10^7 ticks.

quantities. Due to the fact that all investigated datasets clearly exhibited non-linear relationships, we did not only compute R for the entire datasets, but also separately for the parts of the data where the local properties were below or above the average. Our analyses confirmed the order stated in Sect. 5.3.4. Figure 5.15 shows scatterplots of the number of knowledge units N_B with respect to the investigated local properties in A for an example system ($\gamma = 0.75$, $p_a = 0$, $p_k = 0.6$, knowledge exploitation strategy) after 10^7 ticks. The corresponding Pearson correlation coefficients are given in the plot legends.

D) Network-Generating Algorithms

Erdős–Rényi networks: First, a set of N_A isolated (i.e., unconnected) agents is created. Then, from the candidate set of the possible $M_{\max}$ undirected edges that could potentially connect pairs of agents, we choose one random edge, add it to the network, and remove the corresponding edge from the candidate set. The probability of choosing any particular edge is the same for all edges. Then, with the remaining candidate set of $M_{\max} - 1$ edges, we repeat this process until M_A edges are added to the network. If the resulting Erdős–Rényi network contains disconnected components, the network is discarded and the whole process is repeated until a network with only one component is found.

Barabási–Albert networks: A Barabási–Albert network (Barabási and Albert 1999, 2002) is usually created by starting with a connected network of m_0 agents. Then, individual agents are added to the network successively and connected by $m \leq m_0$ nodes until the desired network size is reached, while m is fixed for all added agents. The probability of connecting an added agent to any particular pre-existing agent in the network is weighted by the individual degrees of the pre-existing agents. This procedure is also known as *preferential attachment*. Thus, the resulting network cannot have any disconnected components by construction.

Still, we need a deterministic procedure to derive the values of m and m_0 from the given values N_A and M_A. For determining m, we introduce the convention

$$m = \left\lfloor \frac{M_A}{N_A} \right\rfloor, \tag{5.6}$$

where the *floor* operator $\lfloor \cdot \rfloor$ returns the closest integer number smaller than or equal to its operand. In order to be able to derive a simple expression for m_0, we require the initial network of m_0 agents to be completely connected, i.e., the number of edges is initially equal to $\frac{m_0(m_0-1)}{2}$. By additionally requiring $M_A = kN_A$ with $k \in \mathbb{N}_+$, Eq. (5.6) simplifies to

$$m = \frac{M_A}{N_A}. \tag{5.7}$$

Since the number of agents successively added to this network is $N_A - m_0$, we can now derive m_0 from N_A and M_A:

$$M_A = \underbrace{\frac{m_0(m_0-1)}{2}}_{\text{edges of initial network}} + \underbrace{(N_A - m_0)m}_{\text{edges added by preferential attachment}}$$

$$\stackrel{(5.7)}{=} \frac{1}{2}m_0^2 - \frac{1}{2}m_0 + M_A - \frac{M_A}{N_A}m_0 \qquad | -M_A$$

$$0 = m_0^2 - m_0\left(2\frac{M_A}{N_A} + 1\right) \qquad | \times \frac{1}{m_0}$$

$$= m_0 - \left(2\frac{M_A}{N_A} + 1\right)$$

$$\Rightarrow m_0 = 2\frac{M_A}{N_A} + 1. \tag{5.8}$$

Watts–Strogatz networks: The construction of a Watts–Strogatz network (Watts and Strogatz 1998) is started by generating a regular ring network with N_A agents where each agent is connected to its k next neighbors (in one direction of the ring) with $k = \frac{M_A}{N_A} \in \mathbb{N}_+$. Then, the edges of each agent in the ring are rewired with probability p_{WS} to another randomly chosen agent in the network. In order to achieve small-world properties, the rewiring probability is chosen as $p_{WS} = 0.1$. As in the case of an Erdős–Rényi network, the resulting network might have disconnected components. If this is the case, the network is discarded and the aforementioned procedure is repeated until a network with only one single component is found.

Compatibility-Based Networks First, a set of N_A isolated (i.e., unconnected) agents is created and each agent is assigned an AKU with bit randomization probability p_a. Then, a candidate list E is created containing all $M_{max} = N_A (N_A - 1)$ possible edges with weights assigned according to the pairwise compatibility of the agent's different AKUs. Thereafter, we proceed with the following iterative process with an index n, initialized with $n = 0$:

1. Sort E by weight in descending order and move the first n elements to the end of E.
2. Connect agents by the first M_A edges in E.
3. (a) The network has disconnected components:
   Discard the network and shuffle E randomly.
   If the procedure has been repeated a multiple of N_A times without success, increase n by one.
   If $n \geq M_{max}$, throw an error. Otherwise, proceed with step 1.
   (b) The network is fully connected:
   Exit procedure.

References

Ahrweiler, P., Gilbert, N., & Pyka, A. (Eds.). (2016). *Joining complexity science and social simulation for innovation policy: Agent-based modelling using the SKIN platform*. Newcastle upon Tyne: Cambridge Scholars Publishing.

Ahrweiler, P., & Keane, M. T. (2013). Innovation networks. *Mind and Society, 12*(1), 73–90.

Ancori, B., Bureth, A., & Cohendet, P. (2000). The economics of knowledge: The debate about codification and tacit knowledge. *Industrial and Corporate Change, 9*, 255–287.

Antonelli, C. (2006). The business governance of localized knowledge: An information economics approach for the economics of knowledge. *Industry and Innovation, 13*(3), 227–261.

Antonelli, C., & Link, A. N. (Eds.). (2015). *Routledge handbook of the economics of knowledge*. London: Routledge.

Arthur, W. B. (2007). The structure of invention. *Research Policy, 36*(2), 274–287.

Audretsch, D. B., & Feldman, M. P. (1996). Innovative clusters and the industry life cycle. *Review of Industrial Organization, 11*, 253–273.

Baddeley, M. (2010). Herding, social influence and economic decision-making: Socio-psychological and neuroscientific analyses. *Philosophical Transactions of the Royal Society B, 365*, 281–290.

Baddeley, M. (2013). Herding, social influence and expert opinion. *Journal of Economic Methodology, 20*(1), 35–44.

Barabási, A.-L. (2016). *Network science*. Cambridge: Cambridge University Press.

Barabási, A.-L., & Albert, R. (1999). Emergence of scaling in random networks. *Science, 286*(5439), 509–512.

Barabási, A.-L., & Albert, R. (2002). Statistical mechanics of complex networks. *Reviews of Modern Physics, 74*(1), 47–97.

Barley, W., Treem, J., & Kuhn, T. (2018). Valuing multiple trajectories of knowledge: A critical review and agenda for knowledge management research. *Academy of Management Annals, 12*(1), 278–317.

Barrat, A., Barthélemy, M., & Vespignani, A. (2008). *Dynamical processes on complex networks.* Cambridge: Cambridge University Press.

Bastian, M., Heymann, S., & Jacomy, M. (2009). Gephi: An open source software for exploring and manipulating networks. In *Proceedings of the Third International AAAI Conference on Weblogs and Social Media* (pp. 361–362). Retrieved from http://www.aaai.org/ocs/index.php/ICWSM/09/paper/view/154.

Baum, J. A. C., Cowan, R., & Jonard, N. (2010). Network-independent partner selection and the evolution of innovation networks. *Management Science, 56*(11), 2094–2110.

Blackmore, S. (1999). *The meme machine.* Oxford: Oxford University Press.

Bogner, K., Mueller, M., & Schlaile, M. P. (2018). Knowledge diffusion in formal networks - The roles of degree distribution and cognitive distance. *International Journal of Computational Economics and Econometrics, 8*(3/4), 388–407.

Boschma, R. A. (2005). Proximity and innovation: A critical assessment. *Regional Studies, 39*(1), 61–74.

Boschma, R. A., & Lambooy, J. G. (1999). Evolutionary economics and economic geography. *Journal of Evolutionary Economics, 9*(4), 411–429.

Buchmann, T., & Pyka, A. (2012). Innovation networks. In M. Dietrich & J. Krafft (Eds.), *Handbook on the economics and theory of the firm* (pp. 466–482). Cheltenham: Edward Elgar.

Buskes, C. (1998). *The genealogy of knowledge: A Darwinian approach to epistemology and philosophy of science.* Tilburg: Tilburg University Press.

Buskes, C. (2010). Das Prinzip Evolution und seine Konsequenzen für die Epistemologie und Erkenntnisphilosophie. In M. Delgado, O. Krüger, & G. Vergauwen (Eds.), *Das Prinzip Evolution* (pp. 177–192). Stuttgart: Kohlhammer.

Canals, A. (2005). Knowledge diffusion and complex networks: A model of high-tech geographical industrial clusters. In *Proceedings of the 6th European Conference on Organizational Knowledge, Learning, and Capabilities* (pp. 1–21). Retrieved from https://warwick.ac.uk/fac/soc/wbs/conf/olkc/archive/oklc6/papers/canals.pdf.

Canals, A., Boisot, M., & MacMillan, I. (2008). The spatial dimension of knowledge flows: A simulation approach. *Cambridge Journal of Regions, Economy and Society, 1*(2), 175–204.

Cohen, W. M., & Levinthal, D. A. (1990). Absorptive capacity: A new perspective on learning and innovation. *Administrative Science Quarterly, 35*(1), 128–152.

Collier, N., & North, M. (2013). Parallel agent-based simulation with repast for high performance computing. *Simulation, 89*(10), 1215–1235.

Cowan, R., David, P. A., & Foray, D. (2000). The explicit economics of knowledge codification and tactness. *Industrial and Corporate Change, 9*(2), 211–253.

Cowan, R., & Jonard, N. (2004). Network structure and the diffusion of knowledge. *Journal of Economic Dynamics and Control, 28*, 1557–1575.

Cowan, R., & Jonard, N. (2007). Structural holes, innovation and the distribution of ideas. *Journal of Economic Interaction and Coordination, 2*(2), 93–110.

Cowan, R., & Jonard, N. (2009). Knowledge portfolios and the organization of innovation networks. *The Academy of Management Review, 34*(2), 320–342.

Cowan, R., Jonard, N., & Zimmermann, J.-B. (2006). Evolving networks of inventors. *Journal of Evolutionary Economics, 16*(1–2), 155–174.

Crawford, M. B. (2015). *The world beyond your head: On becoming an individual in an age of distraction.* New York: Farrar, Straus and Giroux.

Davenport, T. H., & Beck, J. C. (2001). *The attention economy: Understanding the new currency of business.* Boston: Harvard Business School Press.

Dennett, D. C. (1995). *Darwin's dangerous idea: Evolution and the meanings of life.* London: Simon & Schuster.

Dennett, D. C. (2017). *From bacteria to Bach and back: The evolution of minds*. New York: W. W. Norton.
Distin, K. (2005). *The selfish meme. A critical reassessment*. Cambridge: Cambridge University Press.
Dopfer, K. (2012). The origins of meso economics: Schumpeter's legacy and beyond. *Journal of Evolutionary Economics, 22*(1), 133–160.
Dopfer, K., Foster, J., & Potts, J. (2004). Micro-meso-macro. *Journal of Evolutionary Economics, 14*, 263–279.
Dopfer, K., & Potts, J. (2008). *The general theory of economic evolution*. London: Routledge.
Dosi, G. (1988). The nature of the innovative process. In G. Dosi, C. Freeman, R. Nelson, G. Silverberg, & L. Soete (Eds.), *Technical change and economic theory* (pp. 221–238). London: Pinter Publishers.
Dosi, G., Fagiolo, G., & Marengo, L. (2001). On the dynamics of cognition and actions. An assessment of some models of learning and evolution. In A. Nicita & U. Pagano (Eds.), *The evolution of economic diversity* (pp. 164–196). London: Routledge.
Egbetokun, A., & Savin, I. (2014). Absorptive capacity and innovation: When is it better to cooperate? *Journal of Evolutionary Economics, 24*(2), 399–420.
Erdős, P., & Rényi, A. (1959). On random graphs. *Publicationes Mathematicae, 6*, 290–297.
Erdős, P., & Rényi, A. (1960). On the evolution of random graphs. *A Matematikai Kutató Intézet Közleményei, 5*(A1–2), 17–61.
Falkinger, J. (2007). Attention economies. *Journal of Economic Theory, 133*, 266–294.
Falkinger, J. (2008). Limited attention as a scarce resource in information-rich economies. *The Economic Journal, 118*(532), 1596–1620.
Feldman, M. F., & Audretsch, D. B. (1999). Innovation in cities: Science-based diversity, specialization and localized competition. *European Economic Review, 43*, 409–429.
Ferrari, D., Read, D., & van der Leeuw, S. (2009). An agent-based model of information flows in social dynamics. In D. Lane, S. van der Leeuw, D. Pumain, & G. West (Eds.), *Complexity perspectives in innovation and social change* (pp. 389–412). Dordrecht: Springer.
Foray, D. (2004). *Economics of knowledge*. Cambridge, MA: The MIT Press.
Foray, D. (2014). *Smart specialisation: Opportunities and challenges for regional innovation policy*. London: Routledge.
Foray, D., & Mairesse, J. (2002). The knowledge dilemma and the geography of innovation. In M. P. Feldman & N. Massard (Eds.), *Institutions and systems in the geography of innovation* (pp. 35–54). New York: Springer.
Francisco, A. P., & Oliveira, A. L. (2011). On community detection in very large networks. In L. Costa, A. Evsukoff, G. Mangioni, & R. Menezes (Eds.), *Complex networks: Second international workshop, CompleNet 2010* (pp. 208–216). Heidelberg: Springer.
Frenken, K., van Oort, F., & Verburg, T. (2007). Related variety, unrelated variety and regional economic growth. *Regional Studies, 41*(5), 685–697.
Frieze, A., & Karoński, M. (2016). *Introduction to random graphs*. Cambridge: Cambridge University Press.
Galassi, M., Davies, J., Thelier, J., Gough, B., Jungman, G., Alken, P., & Rossi, F. (2009). *GNU Scientific Library reference manual* (3rd ed.). Network Theory Limited.
Garcia, R. (2005). Uses of agent-based modeling in innovation/new product development research. *The Journal of Product Innovation Management, 22*, 380–398.
Gilbert, E. N. (1959). Random graphs. *Annals of Mathematical Statistics, 30*(4), 1141–1144.
Gilbert, N., Ahrweiler, P., & Pyka, A. (2007). Learning in innovation networks: Some simulation experiments. *Physica A: Statistical Mechanics and its Applications, 378*, 100–109.
Gilbert, N. (2008). *Agent-based models*. Thousand Oaks: Sage.
Gilbert, N., Ahrweiler, P., & Pyka, A. (Eds.). (2014). *Simulating knowledge dynamics in innovation networks*. Berlin: Springer.
Gross, T., & Blasius, B. (2008). Adaptive coevolutionary networks: A review. *Journal of the Royal Society Interface, 5*(20), 259–271.

Gupta, Y., Saxena, A., Das, D., & Iyengar, S. R. S. (2016). Modeling memetics using edge diversity. In H. Cherifi, B. Gonçalves, R. Menezes, & R. Sinatra (Eds.), *Complex networks VII. proceedings of the 7th workshop on complex networks complenet 2016* (pp. 187-198). Cham: Springer.

Halford, G. S., Wilson, W. H., & Phillips, S. (2010). Relational knowledge: The foundation of higher cognition. *Trends in Cognitive Sciences, 14*(11), 497–505.

Hamill, L., & Gilbert, N. (2016). *Agent-based modelling in economics*. Chichester: Wiley.

Hamming, R. W. (1950). Error detecting and error correcting codes. *Bell Labs Technical Journal, 29*(2), 147–160.

Hayek, F. A. (1952). *The sensory order: An inquiry into the foundations of theoretical psychology*. Chicago: The University of Chicago Press.

Heylighen, F., & Chielens, K. (2009). Evolution of culture, memetics. In R. A. Meyers (Ed.), *Encyclopedia of complexity and systems science* (pp. 3205–3220). New York: Springer.

Hodgson, G. M., & Knudsen, T. (2010). *Darwin's conjecture: The search for general principles of social and economic evolution*. Chicago: University of Chicago Press.

Hodgson, G. M., & Knudsen, T. (2012). Agreeing on generalised Darwinism: A response to Pavel Pelikan. *Journal of Evolutionary Economics, 22*(1), 9–18.

Jackson, M. O., & Yariv, L. (2011). Diffusion, strategic interaction, and social structure. In J. Benhabib, A. Bisin, & M. O. Jackson (Eds.), *Handbook of social economics* (Vol. 1A, pp. 645–678). Amsterdam: Elsevier.

Jensen, M. B., Johnson, B., Lorenz, E., & Lundvall, B.-Å. (2007). Forms of knowledge and modes of innovation. *Research Policy, 36*(5), 680–693.

Kiesling, E., Günther, M., Stummer, C., & Wakolbinger, L. M. (2012). Agent-based simulation of innovation diffusion: A review. *Central European Journal of Operations Research, 20*(2), 183–230.

Klarl, T. A. (2014). Knowledge diffusion and knowledge transfer revisited: Two sides of the medal. *Journal of Evolutionary Economics, 24*(4), 737–760.

Klein, M., & Sauer, A. (2016). Celebrating 30 years of innovation system research: What you need to know about innovation systems. *Hohenheim Discussion Papers in Business, Economics and Social Sciences*, 17-2016. Retrieved from http://nbn-resolving.de/urn:nbn:de:bsz:100-opus-12872.

Koschatzky, K., Kulicke, M., & Zenker, A. (Eds.). (2001). *Innovation networks: Concepts and challenges in the European perspective*. Berlin: Springer.

Kuhn, T. S. (1996). *The structure of scientific revolutions* (3rd ed.). Chicago: The University of Chicago Press.

Lamberson, P. J. (2016). Diffusion in networks. In Y. Bramoullé, A. Galeotti, & B. W. Rogers (Eds.), *The Oxford handbook of the economics of networks* (pp. 479–503). Oxford: Oxford University Press.

Langrish, J. Z. (2017). Physics or biology as models for the study of innovation. In B. Godin & D. Vinck (Eds.), *Critical studies of innovation: Alternative approaches to the pro-innovation bias* (pp. 296–318). Cheltenham: Edward Elgar.

Leonard, D. A. (2006). Innovation as a knowledge generation and transfer process. In A. Singhal & J. W. Dearing (Eds.), *Communication of innovations: A journey with Ev Rogers* (pp. 83–110). New Delhi: Sage.

Lerman, K. (2016). Information is not a virus, and other consequences of human cognitive limits. *Future Internet, 8*(2). https://doi.org/10.3390/fi8020021.

Lundvall, B.-Å. (2004). The economics of knowledge and learning. In J. L. Christensen & B.-Å. Lundvall (Eds.), *Product innovation, interactive learning and economic performance (research on technological innovation, management and policy, volume 8)* (pp. 21–42). Amsterdam: Elsevier.

Lundvall, B.-Å. (2016). *The learning economy and the economics of hope*. London: Anthem.

Lundvall, B.-Å., & Johnson, B. (1994). The learning economy. *Journal of Industry Studies, 1*(2), 23–42.

Luo, S., Du, Y., Liu, P., Xuan, Z., & Wan, Y. (2015). A study on coevolutionary dynamics of knowledge diffusion and social network structure. *Expert Systems with Applications, 42*(7), 3619–3633.

March, J. G. (1991). Exploration and exploitation in organizational learning. *Organization Science, 2*(1), 71–87.

Markey-Towler, B. (2016). *Foundations for economic analysis: The architecture of socioeconomic complexity* (Doctoral dissertation, The University of Queensland, School of Economics).

Markey-Towler, B. (2017). *Narratives and Chinese Whispers: Ideas and knowledge in bubbles, diffusion of technology and policy transmission*. Paper presented at the 10th European Meeting on Applied Evolutionary Economics, May 31 - June 3, 2017, in Strasbourg. https://doi.org/10.2139/ssrn.2912739.

Matsumoto, M., & Nishimura, T. (1998). Mersenne twister: A 623-dimensionally equidistributed uniform pseudo-random number generator. *ACM Transactions on Modeling and Computer Simulation, 8*(1), 3–30.

Mokyr, J. (1998). Science, technology, and knowledge: What historians can learn from an evolutionary approach. *Papers on Economics and Evolution*, No. 9803, Max-Planck-Institute for Research into Economic Systems, Jena.

Mokyr, J. (2002). *The gifts of Athena: Historical origins of the knowledge economy*. Princeton: Princeton University Press.

Mokyr, J. (2017). *A culture of growth: The origins of the modern economy - The Graz Schumpeter lectures*. Princeton: Princeton University Press.

Morone, A., Morone, P., & Taylor, R. (2007). A laboratory experiment of knowledge diffusion dynamics. In U. Cantner & F. Malerba (Eds.), *Innovation, industrial dynamics and structural transformation: Schumpeterian legacies* (pp. 283–302). Berlin: Springer.

Morone, P., & Taylor, R. (2004). Knowledge diffusion dynamics and network properties of face-to-face interactions. *Journal of Evolutionary Economics, 14*(3), 327–351.

Morone, P., & Taylor, R. (2009). Knowledge architecture and knowledge flows. In M. Khosrow-Pour (Ed.), *Encyclopedia of information science and technology* (2nd ed., pp. 2319–2324). Hershey: IGI Global.

Morone, P., & Taylor, R. (2010). *Knowledge diffusion and innovation: Modelling complex entrepreneurial behaviours*. Cheltenham: Edward Elgar.

Mueller, M., Schrempf, B., & Pyka, A. (2015). Simulating demand-side effects on innovation. *International Journal of Computational Economics and Econometrics, 5*(3), 220–236.

Mueller, M., Bogner, K., Buchmann, T., & Kudic, M. (2017). The effect of structural disparities on knowledge diffusion in networks: An agent-based simulation model. *Journal of Economic Interaction and Coordination, 12*(3), 613–634.

Müller, M. (2017). *An agent-based model of heterogeneous demand*. Wiesbaden: Springer.

Müller, M., Buchmann, T., & Kudic, M. (2014). Micro strategies and macro patterns in the evolution of innovation networks: An agent-based simulation approach. In N. Gilbert, P. Ahrweiler, & A. Pyka (Eds.), *Simulating knowledge dynamics in innovation networks* (pp. 73–95). Berlin: Springer.

Namatame, A., & Chen, S.-H. (2016). *Agent-based modeling and network dynamics*. Oxford: Oxford University Press.

Newman, M. E. J. (2004a). Analysis of weighted networks. *Physical Review E, 70*(056131). https://doi.org/10.1103/PhysRevE.70.056131.

Newman, M. E. J. (2004b). Fast algorithm for detecting community structure in networks. *Physical Review E, 69*(066133). https://doi.org/10.1103/PhysRevE.69.066133.

Newman, M. E. J., & Girvan, M. (2004). Finding and evaluating community structure in networks. *Physical Review E, 69*(026113). https://doi.org/10.1103/PhysRevE.69.026113.

Newman, M. E. J. (2010). *Networks: An introduction*. Oxford: Oxford University Press.

Nooteboom, B. (1999). *Inter-firm alliances: Analysis and design*. London: Routledge.

Nooteboom, B. (2009). *A cognitive theory of the firm: Learning, governance and dynamic capabilities*. Cheltenham: Edward Elgar.

Nooteboom, B., Van Haverbeke, W., Duysters, G., Gilsing, V., & van den Oord, A. (2007). Optimal cognitive distance and absorptive capacity. *Research Policy, 36*(7), 1016–1034.

Polanyi, M. (1966). *The tacit dimension. With a new foreword by Amartya Sen* (revised 2009 edn.). Chicago: The University of Chicago Press.

Pyka, A., & Küppers, G. (Eds.). (2002). *Innovation networks: Theory and practice*. Cheltenham: Edward Elgar.

Reagans, R., & McEvily, B. (2003). Network structure and knowledge transfer: The effects of cohesion and range. *Administrative Science Quarterly, 48*(2), 240–267.

Rizzello, S. (2004). Knowledge as a path-dependence process. *Journal of Bioeconomics, 6*(3), 255–274.

Rogers, E. M. (2003). *Diffusion of innovations* (5th ed.). New York: Simon and Schuster.

Roy, D. (2017). Myths about memes. *Journal of Bioeconomics, 19*(3), 281–305.

Sackmann, S. A. (1991). *Cultural knowledge in organizations: Exploring the collective mind*. Newbury Park, CA: Sage.

Savin, I., & Egbetokun, A. (2016). Emergence of innovation networks from R & D cooperation with endogenous absorptive capacity. *Journal of Economic Dynamics and Control, 64*, 82–103.

Saviotti, P. P. (2009). Knowledge networks: Structure and dynamics. In A. Pyka & A. Scharnhorst (Eds.), *Innovation networks: New approaches in modelling and analyzing* (pp. 19–41). Berlin: Springer.

Saviotti, P. P. (2011). Knowledge, complexity and networks. In C. Antonelli (Ed.), *Handbook on the economic complexity of technological change* (pp. 141–180). Cheltenham: Edward Elgar.

Schlaile, M. P., & Ehrenberger, M. (2016). Complexity, cultural evolution, and the discovery and creation of (social) entrepreneurial opportunities: Exploring a memetic approach. In E. S. C. Berger & A. Kuckertz (Eds.), *Complexity in entrepreneurship, innovation and technology research: Applications of emergent and neglected methods* (pp. 63–92). Cham: Springer.

Schlaile, M. P., Mueller, M., Schramm, M., & Pyka, A. (2018). Evolutionary economics, responsible innovation and demand: Making a case for the role of consumers. *Philosophy of Management, 17*(1), 7–39.

Schlaile, M. P. (2018). A case for (econo-)memetics: Why we should not throw the baby out with the bathwater. Presented at The Generalized Theory of Evolution conference, January 31st to February 3rd, 2018 in Duesseldorf.

Schmid, S. (2015). *Organizational learning in innovation networks: Exploring the role of cognitive distance and absorptive capacity - an agent-based model*. Marburg: Metropolis.

Siek, J. G., Lee, L.-Q., & Lumsdaine, A. (2002). *The Boost graph library: User guide and reference manual*. Upper Saddle River, NJ: Pearson Education.

Simon, H. A. (1971). Designing organizations for an information-rich world. In M. Greenberger (Ed.), *Computers, communication, and the public interest* (pp. 37–72). Baltimore, MD: Johns Hopkins Press.

Smith, K. (2000). What is the 'knowledge economy'? Knowledge-intensive industries and distributed knowledge bases. Prepared as part of the project "Innovation Policy in a Knowledge-Based Economy" commissioned by the European Commission, presented at the DRUID Summer Conference, Aalborg, Denmark, June 2000. Retrieved from https://www.knowledge4all.com/Temp/Files/95c9162b-b420-4b49-9b2b-5a7bac1c5539.pdf.

Sobolevsky, S., Campari, R., Belyi, A., & Ratti, C. (2014). General optimization technique for high-quality community detection in complex networks. *Physical Review E, 90*(012811). https://doi.org/10.1103/PhysRevE.90.012811.

Speel, H.-C. (1999). Memetics: On a conceptual framework for cultural evolution. In F. Heylighen, J. Bollen, & A. Riegler (Eds.), *The evolution of complexity: The violet book of "Einstein meets Magritte"* (pp. 229–254). Dordrecht: Kluwer Academic Publishers.

Spitzberg, B. H. (2014). Toward a model of meme diffusion (M^3D). *Communication Theory, 24*(3), 311–339.

Szulanski, G. (2003). *Sticky knowledge: Barriers to knowing in the firm*. London: Sage.

Tur, E. M., & Azagra-Caro, J. M. (2018). The coevolution of endogenous knowledge networks and knowledge creation. *Journal of Economic Behavior and Organization, 145*, 424–434.

Tur, E. M., Zeppini, P., & Frenken, K. (2014). Diffusion of ideas, social reinforcement and percolation. In *Social simulation conference*. Autónoma University of Barcelona. Retrieved from https://ddd.uab.cat/pub/poncom/2014/128046/ssc14_a2014a41iENG.pdf.

Tur, E. M., Zeppini, P., & Frenken, K. (2018). Diffusion with social reinforcement: The role of individual preferences. *Physical Review E, 97*. https://doi.org/10.1103/PhysRevE.97.022302.

Tywoniak, S. A. (2007). Knowledge in four deformation dimensions. *Organization, 14*(1), 53–76.

Valente, T. W. (2006). Communication network analysis and the diffusion of innovations. In A. Singhal & J. W. Dearing (Eds.), *Communication of innovations: A journey with Ev Rogers* (pp. 61–82). New Delhi: Sage.

Vermeulen, B., & Pyka, A. (2017). The role of network topology and the spatial distribution and structure of knowledge in regional innovation policy: A calibrated agent-based model study. *Computational Economics*. https://doi.org/10.1007/s10614-017-9776-3.

von Bülow, C. (2013). Meme. [English translation of the (German) article "Mem". In J. Mittelstraß (Ed.), *Enzyklopädie Philosophie und Wissenschaftstheorie* (2nd ed., Vol. 5, pp. 318–324). Stuttgart: Metzler]. Retrieved from https://www.philosophie.uni-konstanz.de/typo3temp/secure_downloads/87495/0/de0f56268a8ad66b13cfc7652e092ce47ea79fb6/meme.pdf.

von Hippel, E. (1994). "Sticky information" and the locus of problem solving: Implications for innovation. *Management Science, 40*(4), 429–439.

Wasserman, S., & Faust, K. (1994). *Social network analysis: Methods and applications*. Cambridge: Cambridge University Press.

Watts, D. J., & Strogatz, S. H. (1998). Collective dynamics of 'small-world' networks. *Nature, 393*, 440–442.

Weng, L. (2014). *Information diffusion on online social networks* (Doctoral dissertation. Retrieved from: School of Informatics and Computing, Indiana University). Retrieved from http://lilianweng.github.io/papers/weng-thesis-single.pdf.

Weng, L., Flammini, A., Vespignani, A., & Menczer, F. (2012). Competition among memes in a world with limited attention. *Scientific Reports, 2*, 335. https://doi.org/10.1038/srep00335.

Weng, L., Menczer, F., & Ahn, Y.-Y. (2013). Virality prediction and community structure in social networks. *Scientific Reports, 3*, https://doi.org/10.1038/srep02522.

Wersching, K. (2010). Schumpeterian competition, technological regimes and learning through knowledge spillover. *Journal of Economic Behavior and Organization, 75*(3), 482–493.

Wilensky, U., & Rand, W. (2015). *An introduction to agent-based modeling: Modeling natural, social, and engineered complex systems with NetLogo*. Cambridge, MA: MIT Press.

Wuyts, S., Colombo, M. G., Dutta, S., & Nooteboom, B. (2005). Empirical tests of optimal cognitive distance. *Journal of Economic Behavior and Organization, 58*(2), 277–302.

Zirulia, L. (2012). Book review: Piergiuseppe Morone and Richard Taylor: Knowledge diffusion and innovation: Modelling complex entrepreneurial behaviours. *Journal of Evolutionary Economics, 22*(2), 395–400.

Chapter 6
Viral Ice Buckets: A Memetic Perspective on the ALS Ice Bucket Challenge's Diffusion

Michael P. Schlaile, Theresa Knausberg, Matthias Mueller, and Johannes Zeman

Abstract This paper presents an exploratory memetic perspective on the diffusion pattern of the ALS Ice Bucket Challenge. More precisely, the paper contributes to research on social learning, cultural evolution, and social contagion by shedding light on endogenous (meme-related) as well as exogenous (structural) properties that may have influenced the Ice Bucket Challenge's diffusion. In a first pillar, we present a descriptive memetic analysis of the diffusion pattern, including an evaluation of the Ice Bucket Challenge according to memetic criteria for successful replication. In the second pillar, we present an agent-based simulation model designed to illuminate the influence of particular social network characteristics on the Ice Bucket Challenge's diffusion. By combining these two pillars, we contribute to the advancement of memetic theory, narrowing the gap between a solely meme-centered perspective and social network analysis.

This chapter has been previously published with open access and should be cited as Schlaile, M.P., Knausberg, T., Mueller, M., & Zeman, J. (2018). Viral ice buckets: A memetic perspective on the ALS Ice Bucket Challenge's diffusion. *Cognitive Systems Research, 52*, 947–969. doi: https://doi.org/10.1016/j.cogsys.2018.09.012. The authors benefited from presenting an earlier draft of this paper at the European Academy of Management Conference, June 21–24, 2017, in Glasgow, Scotland. We are grateful to two anonymous reviewers and the participants of our session, most notably Dermot Breslin and Stephen Dobson, for helpful comments, suggestions, and constructive criticism. All remaining errors and omissions remain our responsibility. M.M. gratefully acknowledges financial support from the Dieter Schwarz Foundation, and J.Z. from the Deutsche Forschungsgemeinschaft (DFG) through the cluster of excellence Simulation Technology.

M. P. Schlaile (✉) · M. Mueller
Department of Innovation Economics, University of Hohenheim, Wollgrasweg 23, 70599 Stuttgart, Germany
e-mail: schlaile@uni-hohenheim.de

M. P. Schlaile
Center for Applied Cultural Evolution, 1776 Millrace Drive, Eugene, OR 97403, USA

T. Knausberg
Alumna, University of Hohenheim, Stuttgart, Germany

J. Zeman
Institute for Computational Physics (ICP), University of Stuttgart, Allmandring 3, 70569 Stuttgart, Germany

© The Author(s), under exclusive license to Springer Nature Switzerland AG 2021
M. P. Schlaile (ed.), *Memetics and Evolutionary Economics*, Economic Complexity and Evolution, https://doi.org/10.1007/978-3-030-59955-3_6

6.1 Introduction

During summer 2014, Internet users—particularly those of social media platforms—could hardly avoid seeing videos of people dumping cold water on their heads.[1] *Amyotrophic lateral sclerosis* (ALS), a previously relatively unknown neurodegenerative disorder affecting motor neurons, suddenly gained rapid attention thanks to the viral Internet phenomenon called *ALS Ice Bucket Challenge* (henceforth IBC).[2] Due to its success in raising awareness and funds (e.g., see Fawzy 2016; Fullman 2015; Hrastelj and Robertson 2016; Jang et al. 2017; Mahoney 2015; Sohn 2017; Wicks 2014), analyzing the IBC is not only of interest in the context of information diffusion and social learning (e.g., Adamic et al. 2016; Cheng et al. 2018; Hui et al. 2018) but may also be particularly fruitful for isolating success factors for advertising a good cause (in contrast to many other well-known and often quite foolish "viral" challenges that spread across the Internet, such as the *neknomination*, *whaling*, *planking*, or the more recent *tide pod challenge*, where teenagers have filmed themselves biting on or eating hazardous laundry detergent-filled gel capsules; see also Grimmelmann 2018).

In this paper, we utilize the framework of memetics, which has been defined by Heylighen and Chielens as "the theoretical and empirical science that studies the replication, spread and evolution of memes" (Heylighen and Chielens 2009, p. 3205). Although memes are frequently opposed by cultural evolutionary researchers (e.g., see Acerbi 2016; Gabora 2008), in this article we follow authors such as Coscia (2013), Dobson and Sukumar (2018), Milner (2016), Mitchell (2012), and Shifman (2014), who advocate memetics as a valid approach to studying the evolution and diffusion of various Internet phenomena. Nevertheless, we tend to agree with Boudry (2018b) and Boudry and Hofhuis (2018, p. 156) that "[t]here is no need for a new *science of memetics*, in the sense of a unifying and overarching theory of culture" (italics in original), because "[i]f selfish memes explain everything, they explain nothing" (Boudry and Hofhuis 2018, p. 159). In our understanding, memetics is an interdisciplinary endeavor that is concerned with the evolution and diffusion of (semantic) information, and which can build bridges between different strands of literature particularly by offering a common language (see also Schlaile 2018). It is, therefore, reasonable to argue that a memetic perspective can generate insights that are useful also for viral marketing campaigns, be it for commercial or charitable reasons (Burgess et al. 2018; Dobson and Sukumar 2018). On a related note, Aunger (2002, p. 16) claims that "viral marketers even acknowledge, in their humbler moments, that their ideas originate in … books about memes." Taking these considerations into account, we argue that memetics can provide a suitable perspective for the analysis of the IBC's success in raising funds and awareness via social

---

[1] Note that more or less successful attempts were made to revive the phenomenon during subsequent summers.
[2] For the history of the IBC, see http://webgw.alsa.org/site/PageServer?pagename=GW_edau_ibc_history_infographic

media platforms. Still, memetics has often been accused of providing a theoretical framework of biological analogies without sufficient practical application (e.g., see Edmonds 2002, 2005; Hull 2000; Schlaile and Constantinescu 2016). Hence, the central goal of this article is to contribute to research on social learning and cultural transmission by means of a memetically substantiated exploratory analysis of an *actual* diffusion phenomenon. We aim to tackle two distinct but interrelated research questions:

R1: Which endogenous (meme-centered) elements of the IBC meme contributed to its diffusion pattern?

R2: Can we identify structural factors (i.e., network properties) that influenced the diffusion of the IBC meme on social networks?

While the first question can be dealt with by means of a descriptive memetic analysis, the second question will be addressed with an agent-based simulation model. Mostly due to a lack of available data, we need a computational model that allows us to conduct simulation experiments.

In order to adequately address these two questions, the paper is structured as follows: The subsequent Sect. 6.2 starts with the necessary theoretical background and elucidates the suitability of a memetic perspective and the IBC meme's constitutive elements. Thereafter, in Sect. 6.3, we present a descriptive memetic analysis, including the origins as well as examples of selected variants of the IBC meme, details of the actual IBC's diffusion pattern based on empirical data, and an evaluation of the IBC meme's characteristics according to criteria for successful replication. The second pillar of our paper consists of the agent-based simulation part (Sect. 6.4), where we develop our model and present the results of our simulations for different scenarios and parameter settings. In Sect. 6.5, we discuss the findings and limitations of our approach and compile practical implications for "memetic campaigns" promoting a good cause. Finally, in Sect. 6.6, we draw our conclusion and summarize potential avenues for further research.

6.2 Theoretical Background

6.2.1 Successful Memetic Replication and Diffusion

While Dawkins (1976) originally conceived of memes as replicators by analogy with genes, he later also compared the spreading mechanisms and diffusion patterns of memes to viruses (e.g., Dawkins 1993; Goodenough and Dawkins 1994). This latter point of view is, for example, also represented by Lynch's (1996, p. 9) statement that memetics can, in part, be considered as "an *epidemiology* of ideas" (italics in original),[3] which is perfectly in line with Aunger (2002, p. 17), who describes memetics as "the cultural analogue to the study of how disease-causing pathogens diffuse through

---

[3]For a critique of Lynch's (1996) account of memetics, see Marsden (1999).

populations."[4] This so-called *meme-as-germ* perspective is often presented as a counterpart to the (initially promoted) *meme-as-gene* perspective (see also Álvarez 2004, Blute 2005, 2010, on this discussion). Taking into account both points of view, it can be argued that—depending on the phenomenon under consideration—in some respects memes may display similarities to genes while in others they do so to viruses (see also Cronk 1999). In fact, the somewhat arbitrary distinction may be seen as two sides of the same coin, acknowledging that memes contain an informational element of *instruction* or *rule* for doing things (as do genes), which can spread like a virus that may even subvert the interests of its carriers (as is lucidly illustrated by the example of witchcraft beliefs in early modern Europe presented by Boudry and Hofhuis 2018). In the context of "viral" Internet phenomena, which may generally be regarded as instances of *social contagion*,[5] it is also noteworthy that Marsden (1998) argued already two decades ago that social contagion research and memetics can be considered as complementary strands of literature, where a memetic perspective can serve as the theoretical underpinning for the evolution of social contagion phenomena, while social contagion research provides empirical evidence. The present article may thus be considered as an attempt to synergize these branches of research.

Another discussion in the field of memetics relevant for our article concerns the "locus" of the meme. While various authors have promoted some kind of neurological view of memes as replicators *located in the human brain*, others consider memes as (imitable) *information* independent of their neural substrate (see also Boudry, Boudry 2018b; Boudry and Hofhuis 2018; Schlaile and Ehrenberger 2016). These and other ontological as well as terminological ambiguities still render it somewhat difficult to find a viable definition of a meme (see also Hodgson and Knudsen 2010; Spitzberg 2014). Being aware of this difficulty, in this article, we will not confine memes to their materialization as neural substrates and follow Distin (2014, p. 3), who suggests that "memes might be [defined as] units of representational content: cultural information preserved in a representational form that has a potential effect on or through those who acquire it." More precisely, "memes … control our behaviour in response to the information that they carry, which we can link in our own minds to other such representations, and which preserve their content in a way that can be transmitted to other people" (Distin 2014, pp. 3–4).[6]

---

[4]Note, however, that ideas and (fashion) trends have often been compared to viruses also outside the memetics literature, for example, by Berger and Milkman (2012), Gladwell (2000), Røvik (2011), or the contributors in Kirby and Marsden (2006); see also Pressgrove et al. (2018), for a literature review. Another related strand of literature is that of the so-called "cultural virus theory" advanced particularly by Cullen (1993, 2000).

[5]According to Goldstone and Janssen (2005, p. 427), social contagion "is the spread of an entity or influence between individuals in a population via interactions between agents". See also Marsden (1998) for a collection of equally broad definitions.

[6]For a possible categorization of different kinds of memetic representations, we also refer the reader to the *P-I-E-trichotomy* proposed by Schlaile and Ehrenberger (2016), which distinguishes between *p-memes* (primal memes; genuine "bits" of information), *i-memes* (which are the individual *mental* representations of a meme, e.g., as parts of culturally acquired schemata that serve the meme's human carriers as information filters and heuristics), and *e-memes* (i.e., the *environmental* representations of a meme, e.g., in terms of verbalizations or *signifiers*).

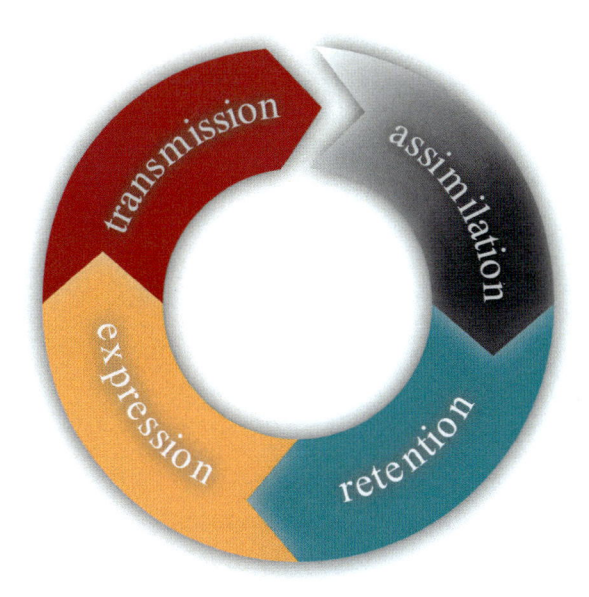

Fig. 6.1 Replication loop according to Heylighen and Chielens (2009). Assimilation of a meme by an individual (carrier) is followed by retention of the meme in that carrier's memory, which, in turn, is followed by expression (e.g., by means of language or behavior), and finally transmission of the meme to other carriers, where the meme (potentially) starts the replication loop all over again. The gap between transmission and assimilation in the figure thus represents the transition of the replicated meme from one carrier to another. Selection takes place at each stage of the loop.

Notably—and in line with the meme-as-germ perspective—pre-existing mental representations could be seen as an analogue to a carrier's immune system, preventing these humans from defenselessly contracting any meme (Aunger 2002). These mental representations (as parts of cultural schemata) may thus serve as information filters and models of the world (Schlaile 2018; Schlaile and Ehrenberger 2016), thereby influencing which and to what extent new memes can be assimilated (Heylighen and Chielens 2009; Schlaile et al. 2018).

According to the related *meme-as-information* perspective (Boudry 2018a; Dennett 1995, 2003, 2017), memes compete for the scarce resource of human attention (Blackmore 1999; Coscia 2014; Distin 2005; Hui et al. 2018; Schlaile 2018; Weng 2014; Weng et al. 2012). In Hofstadter's (1985, p. 51) words: "Various mutations of a meme will have to compete with each other, as well as with other memes, for attention—which is to say, for brain resources in terms of both space and time devoted to that meme." But what is it that makes a meme successful in this competition for attention? Authors such as Heylighen (1993, 1997, 1999) or Spitzberg (2014) have tackled this question and proposed various factors or criteria influencing a meme's selection or replicative success.[7] In this context, it has also been argued that there

---

[7] We will come back to these criteria in Sect. 6.3.3.

exist four distinct but interrelated stages of replication, namely, *assimilation, retention, expression,* and *transmission* (e.g., Chielens and Heylighen 2005; Heylighen and Chielens 2009). These four stages have also been referred to as a "replication loop" (see Fig. 6.1), because the last stage (transmission of the meme) is potentially preceding the assimilation stage in another individual, followed by retention in that individual's memory, then expression, then transmission, and so on (Heylighen and Chielens 2009). Since in this article we are focusing on the *diffusion pattern* of the meme, our discussions below will concentrate mainly on the stages of expression and transmission, albeit without any intention to downplay the importance of the other stages for successful replication and spread of the meme.[8]

6.2.2 The IBC as a Meme

Following Distin's (2014) definition, we can argue that the IBC may be considered a meme: It is preserved in a representational form that can be linked to other representations, it can be transmitted to other individuals, and it contains a set of instructions for behavior. However, it may still be open to debate where this IBC meme ends and where another meme begins. In this regard, it has been purported that memes can cluster together in so-called *co-adapted meme complexes*, often abbreviated to *memeplexes* (e.g., Blackmore 1999; Speel 1999). A memeplex may be understood as a set of memes that replicate better when linked together than on their own. Therefore, it is often difficult to draw clear boundaries between a meme and its memeplex. As Chielens (2003, p. 12) argues, for example, it is possible to basically consider every meme as (part of) a memeplex, as "our own views have been influenced by our own memes and are thus unavoidably linked to them." This aspect is closely related to the *fractal* nature of memes as conceptualized by Velikovsky (2016, 2018) in terms of a so-called *holon/parton structure*. It is quite obvious that the same issue applies to the IBC: Essentially, the IBC can be considered a meme*plex*, consisting of several rule-based elements that may themselves be memes; however, due to its relatively high stability as a spreadable entity, we can also regard the IBC as a discrete unit, despite being made of different meme-components. In line with Rossolatos (2015, p. 133), who argues that the IBC meme is a "multimodal unit … (further decomposable into subunits)", and because we cannot be certain that the IBC memeplex would have succeeded as it did without some of its components, we will henceforth speak of the *IBC meme*, despite being aware that it consists of several distinct elements, which have to be illuminated as well: The IBC meme is mainly a set of rules or instructions, including (a) mentioning the person one was nominated by, (b) responding to the nomination within 24 hours, (c) donating to the ALS Association (or sometimes another ALS-related charity), (d) dumping ice water on one's head, (e) recording and uploading a video of the challenge, (f) nominating three other people to take the challenge up within the next 24 hours, and (g) posting the video while

---

[8] Also note the general discussion about the suitability of the notion of replication in Boudry (2018b).

including one of the related hashtags.[9] Having this list in mind, it becomes evident that the IBC is suitable for being analyzed from a memetic perspective because it is not just a collection of "simple ideas" (e.g., see Dennett 1995, on the distinction between simple and complex ideas). After all, pouring a bucket of ice water over one's head and posting a video of this action online is not exactly an innate activity but complex behavior acquired by imitation.[10] In this context, some authors even claim that the Internet has changed the particularities of memetic evolution, which resulted in an increasing body of academic literature on online memes as a discrete phenomenon (cf. Burgess et al. 2018, p. 3; see also Segev et al. 2015; Wiggins and Bowers 2015, on related discussions, or Sampson 2012, for critical arguments).

Considering the IBC as a meme also prompts us to focus on the meme's characteristics as opposed to the individuals' intentions. Although the IBC meme has not been a genuinely "parasitic" meme subverting the interests of its human carriers (e.g., Boudry 2018a, on a related note), it is still hard to explain its success by invoking *methodological individualism* (e.g., cf. Heath 2015). As Marsden (1998, no pagination) puts it:

> [T]he reason why some social behaviour doesn't seem to make sense from the perspective of the individual is because we are looking at that behaviour at the wrong level. We are taking an anthropocentric or homuncular view of a social world that was created at least in part at a memetic level. Trying to explain the social world from the perspective of the individual is like trying to explain the movements of a car without reference to the driver.

Note, however, that although the so-called *meme's eye view* may suggest that memes can indeed evolve in a direction unforeseen or unintended by their carriers (e.g., Blackmore 2000; Boudry and Hofhuis 2018; Dennett 1995), human influence on the creation and diffusion of memes cannot (and should not) be ruled out either (Boudry 2018a). Sampson (2012, p. 74) boils this issue down to the provocative question: "How can the meme be both omnipotent and manipulable?" Although this philosophical discussion has been and should be addressed elsewhere, we may assume that memetic engineering is—to a certain degree—possible, and that it has not just been the IBC meme's own informational content but also its context and the social structure of the carriers influencing its diffusion.

In the subsequent section, the first pillar of our analysis will, therefore, assume a memetic perspective for describing the IBC meme's diffusion pattern. Although this description should be considered non-exhaustive and perhaps even somewhat fragmented, we deem it necessary for shedding light on the endogenous (i.e., meme-related) factors having contributed to the IBC meme's success. In contrast to merely empirical social contagion research, we may thus provide additional theoretical underpinnings for future "memetic engineering" endeavors. Nevertheless, a merely meme-centered perspective cannot provide sufficient explanation for the IBC meme's

---

[9]For summaries of the IBC's central rules see, e.g., http://www.mtv.com/news/1904680/ice-bucket-challenge-rules/ or Rossolatos (2015).

[10]Remember that Blackmore (1999, p. 43) understands imitation in a broad sense as "passing on information by using language, reading and instruction, as well as other complex skills and behaviours from one person to another."

spreading pattern either. It will, therefore, be necessary to focus also on the characteristics of the social network of carriers as a second pillar of analysis further below (in Sect. 6.4).

6.3 Descriptive Memetic Analysis

6.3.1 Origins and Examples of Variants of the IBC Meme

Despite our article focusing on the diffusion pattern of the IBC meme (i.e., the expression and transmission stages of the replication loop), some initial remarks on the meme's origins and evolution as a differentially selected replicator are advisable. Although it seems impossible to exactly trace the meme's "lineage,"[11] it is plausible to assume that some of its roots lie in the *cold water challenge* and in the *neknomination*, two other Internet phenomena that both arose earlier in the year 2014.[12] The cold water challenge was very popular among associations, such as firefighters, musical societies, or bowling clubs. The challenge counts as successfully completed if the nominated person/club douses itself completely in cold water. People jumped into lakes, danced in swimming pools, or made music in wells, often wearing funny dresses. This meme may have been conducive to inspiring the water element of the IBC.[13] The neknomination prompted nominees to drink an alcoholic beverage in one go and then nominate several other people to do the same (see also Burgess et al. 2018). At least in Germany, where it became known also as the *social beer game*, people were frequently asked to give a crate of beer to the person who nominated them if they refused to accept the challenge. The neknomination, therefore, exhibits a similar nomination structure to that of the IBC and may thus be part of the IBC meme's genealogy. In general, it is not implausible to assume that the cold water challenge and the neknomination (as well as many other viral challenge memes) and, therefore, also the IBC evolved from the meme of the traditional chain letter.

Since around June 2014, a variant of the IBC circulated among a group of US pro athletes, such as golfers and motorcycle racers. At that time, the challenge was already linked to donations for a good cause, but not to a specific kind of charity (Levin 2014). The birth of the *ALS* version of the IBC meme is strongly connected

---

[11] Notably, as Secretan (2013) argues using the example of Picbreeder, memetic evolution may often depend on stigmergic collaborative creation.

[12] E.g., see https://trends.google.com/trends/explore?date=2014-01-01%202014-12-31&q=cold%20water%20challenge,neknomination.

[13] Interestingly, in 2018, the cold water challenge meme went active again among administrations, firefighters, sports clubs, and similar carrier populations. Especially in German-speaking countries like Austria, Germany, and Switzerland it spread as *cold water grill challenge* or, alternatively, *grill pool challenge* (e.g., see https://trends.google.com/trends/explore?q=cold%20water%20grill%20challenge,grill%20pool%20challenge, https://www.youtube.com/results?search_query=cold+water+grill+challenge, or https://www.youtube.com/results?search_query=grill+pool+challenge).

to July 15, 2014.[14] Apart from that, the ALS Association, the largest US non-profit health organization that aims at fighting ALS, mentioned the IBC as a viral movement that spreads awareness for ALS for the first time on August 6, 2014. This day, the ALS Association presented the IBC as inextricably linked to ALS. Hence, the ALS Association had not "invented" the meme, but successfully promoted the combination of the IBC with ALS.

The evolution of the IBC meme did not stop once it was linked to ALS, though. Interestingly, the instructions (mentioned above) usually remained stable at the core, varying only slightly, whereas the performances in the videos sometimes exhibit a high "mutation rate" (see also Rossolatos 2015).[15] Examples include Charlie Sheen dumping $ 10,000 (which he donated) on his head instead of ice water,[16] or Sir Patrick Stewart posting a video of him writing a check and using the ice for a drink.[17]

Soon after the IBC became popular, there also emerged variants that could even be considered as memes *inspired* by the IBC meme instead of mutated versions of it. Examples include the *Rice Bucket Challenge,* which arose in India and involves giving a bucket filled with rice to a nearby needy family.[18] Another one, the *Soapy Water Bucket Challenge* (or: *Lather Against Ebola Challenge*) that originated in Ivory Coast, prompts nominated people to pour a bucket filled with soapy water over their head and give away sanitizer. This challenge shall create awareness for adequate measures of Ebola prevention.[19] Just like the original IBC meme, many of these new memes likewise include the element of raising attention for a good cause.

6.3.2 Remarks on the IBC Meme's Diffusion Pattern

According to Facebook, between June 1 and September 1, 2014, more than 17 million IBC-related videos were shared via Facebook and viewed more than 10 billion times by over 440 million people.[20] Even for those who were not active

---

[14]This day, Chris Kennedy, a professional golfer from Florida, was nominated by a friend to accept the IBC. Chris Kennedy's relative, Anthony Senerchia, suffered from ALS (deceased 2017), which arguably motivated Kennedy to ask the people he nominated to donate to fight ALS. From this day onwards, more and more people associated the IBC with ALS, supported by prominent sufferers from ALS, who helped to promote the combination of these elements.

[15]Chris Kennedy's video was relatively unspectacular: he just poured a standard-sized bucket of ice water over his head, wearing casual clothing, standing in a backyard. Aside from other rather unspectacular videos, later videos showed people using a whole excavator shovel full of water instead of an ice bucket (https://www.youtube.com/watch?v=v49Nblyz4ZA), or pouring liquid nitrogen (https://www.youtube.com/watch?v=peVgAzq-1LQ) over themselves.

[16]https://www.youtube.com/watch?v=qat9gR5nrpM.

[17]https://www.youtube.com/watch?v=wkO4NIqAMss.

[18]http://www.npr.org/blogs/goatsandsoda/2014/08/27/343498667/rice-bucket-challenge-put-rice-in-bucket-do-not-pour-over-head.

[19]http://news.nationalpost.com/2014/08/27/africas-answer-to-ice-bucket-challenge-the-lather-against-ebola-campaign/.

[20]http://newsroom.fb.com/news/2014/08/the-ice-bucket-challenge-on-facebook/.

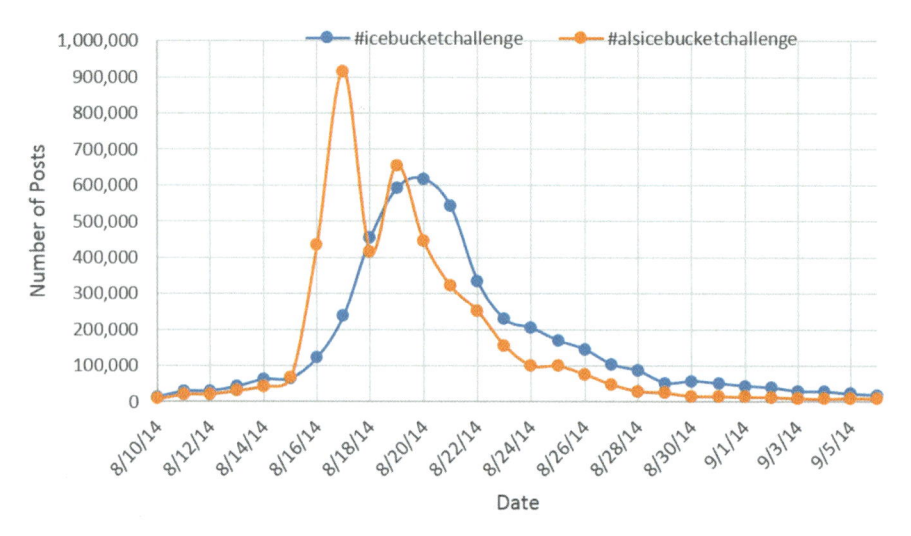

Fig. 6.2 Use of IBC-related hashtags on Twitter (Data Source: Topsy).

on Twitter, Facebook, or other platforms during the summer of 2014, there was still enough possibility to "contract" the IBC meme via traditional media such as newspapers and TV reports. However, the main vectors for the IBC meme were the social media platforms (or social networking sites) of Facebook and Twitter.

So-called hashtags are very helpful for tracking the spread of memes on the Internet. Searching for a certain hashtag on social media platforms allows exploring all posts that were published using this tag. These hashtags have also played a central role in the diffusion of the IBC meme: On their website, the ALS Association recommended using the hashtags *#icebucketchallenge, #alsicebucketchallenge,* and *#strikeoutals* when posting statements related to the IBC. Although not every carrier has used one of these hashtags while posting IBC-related videos via social media platforms, these hashtags were frequently used in the context of the IBC. For tracking hashtags, Twitter has proven more useful than Facebook. The main reason for this is that a large part of Facebook users' posts can only be viewed by their own Facebook friends.

The fact that the IBC meme spread across several different platforms that operate independently of each other renders it very difficult to gather figures about how many people have actually participated in the challenge at a certain point in time. Nevertheless, observing how often the most popular hashtags for the IBC were used at a particular time can serve as an indicator for the diffusion of the meme.

Figure 6.2 shows how many tweets mentioned the hashtags #icebucketchallenge or #alsicebucketchallenge each day between August 10 and September 6, 2014.[21] Although both hashtags did not evolve simultaneously, their use increased rapidly around August 15. The hashtag #alsicebucketchallenge peaked on August 17 (used

---

[21] Data were gathered using Topsy shortly before the service was shut down by Apple.

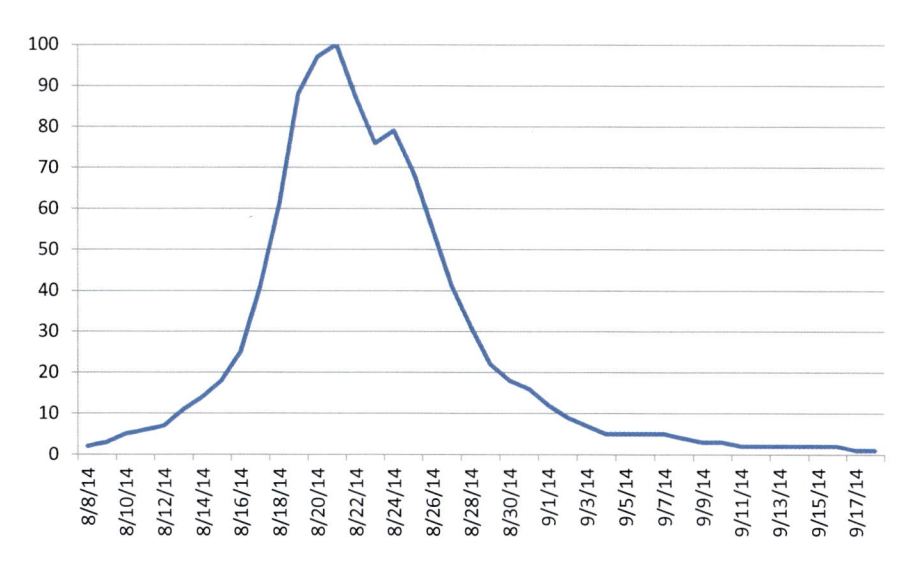

Fig. 6.3 Worldwide Google search interest (popularity) for "ALS Ice Bucket Challenge" (Data Source: Google Trends).

in 915,491 tweets on this day) and, after a brief decline, peaked again on August 19, mentioned in 665,825 tweets that day. In comparison, #icebucketchallenge increased more slowly than its counterpart with the ALS prefix and peaked on August 20, when it was used in 617,305 tweets. The numbers and curves show the immense popularity of the IBC meme on Twitter while at the same time underlining its short lifespan.

Another way to reconstruct the spreading pattern of the IBC meme over time is the search volume for the term "ALS Ice Bucket Challenge" on Google, the most widely used Internet search engine. Figure 6.3 uses data from Google Trends to provide a visualization of this search term's popularity. Google Trends divides each data point by the total searches of the geography and time range it represents to compare relative popularity. The resulting numbers are scaled on a range of 0–100 based on the search term's proportion to all searches on all topics (in that region).[22] After its take-off around August 6, the search interest rose steeply. On August 21, it already reached its peak. This sharp rise is in line with the increase of Twitter posts referring to the IBC during this period. After the peak on August 21, a steep decline in search interest occurs. A second peak can be seen on August 24 but starting August 25 it decreased constantly.

By taking a closer look at the regional origins of the search term popularities, we may explain the second, smaller peak observable for both the #alsicebucketchallenge hashtag and the worldwide Google searches: Fig. 6.4 depicts the (relative) popularities of searches for "ALS ice bucket challenge" in Canada, the United States of America (USA), Germany, Russia, and the United Kingdom (UK). We decided to use Canada as a direct neighbor of the USA (both geographically and in terms of

---

[22] https://support.google.com/trends/answer/4365533?hl=en&ref_topic=6248052.

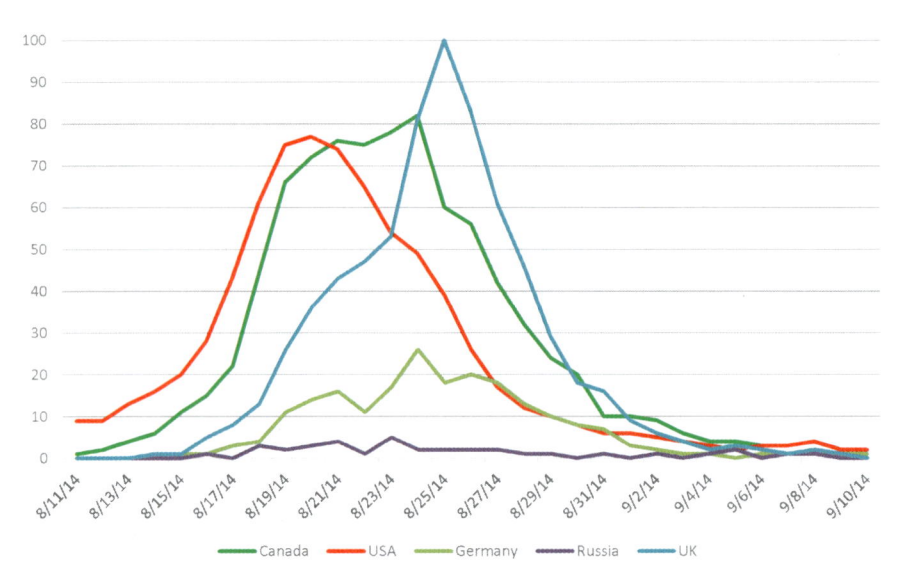

Fig. 6.4 Google search interest (popularity) of "ALS ice bucket challenge" compared between Canada, USA, Germany, Russia, and the UK (Google Trends).

language overlaps), Germany, and Russia as examples for more geographically and culturally distant (particularly in terms of language) carrier populations, and the UK as a geographically distant population with a common language.

As Fig. 6.4 suggests, the second peak is likely to be the result of a later reception of the IBC meme in countries geographically (and culturally) distant from the USA. Moreover, although not exactly representative for culturally distant populations in general, the relatively low popularity of the search interest in Germany and Russia may imply that part of the IBC meme's success may have been the compatibility with the culture of anglophone countries.

The double-peaked diffusion pattern is also supported by the shape of the curve of daily donations to the ALS Association between August 13 and August 29, 2014, albeit with a certain delay, as depicted in Fig. 6.5.[23]

Nevertheless, as noted above, it is impossible to observe how many people actually contracted the meme at certain points in time, particularly due to a lack of available data. However, the curves indicate a comparable time span during which the meme spread successfully: the period between August 10 and September 10, 2014. However, it should be noted that the massive decline of both the search interest for "ALS Ice Bucket Challenge" and the use of related hashtags does not imply that the IBC meme has ceased to exist. It just indicates that most carriers who adopted the meme do not actively express it any more (by using related hashtags) and are less active

---

[23] After August 29, the ALS Association stopped reporting the amount of daily donations. However, on September 22, 2014, they announced that they had received an impressive $ 115 million in IBC donations.

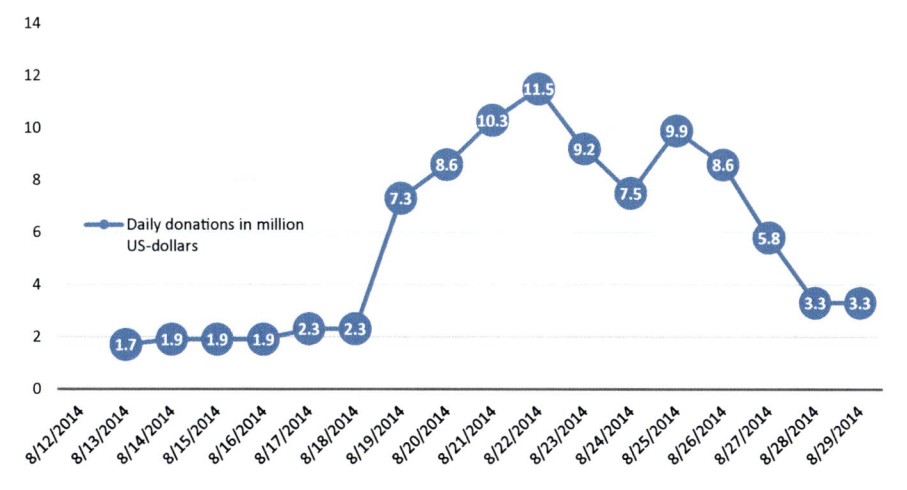

Fig. 6.5 Daily donations to the ALS Association (Data Source: ALS Association).

in reinforcing the meme by gathering more information about it. Hence, the rate of new IBC meme contagions can be assumed to have declined accordingly.

6.3.3 Important Characteristics for the IBC's Diffusion

Several authors have proposed factors or criteria that contribute to a meme's successful replication, most notably Heylighen (1993, 1997, 1999), Chielens and Heylighen (2005), Heylighen and Chielens (2009), and Spitzberg (2014). Drawing eclectically on these works, in this section, we highlight characteristics that we may consider to be central for the IBC meme's diffusion.[24] The evaluation commences with (a) the *meme level*, then moves to (b) the *level of the individual (human) carrier*, and concludes with the population of carriers particularly in terms of (c) the *network level*.

At the *meme level* (a), the diffusion of the IBC meme can be assumed to have been positively influenced by its perceptibility via various media formats (Spitzberg 2014: *media convergence*; Heylighen, 1999: *invariance*); the IBC meme was transmitted via different (social) media platforms, including YouTube and (online) newspapers. Its diffusion through different formats was, to a certain extent, limited by the fact that the meme is closely linked to its visualization, which makes it predominantly suitable for visual media formats. Nonetheless, the meme's presence across different types of media implies that many (potential) carriers were confronted with it repeatedly,

---

[24] Note, however, that this evaluation consequently relies strongly on the authors' subjective assessment.

which surely had a positive influence on the assimilation and retention of the meme (Heylighen and Chielens 2009: *repetition*).

As simpler memes tend to be adopted and expressed more easily (Heylighen and Chielens 2009; Spitzberg 2014: *simplicity*), it is advantageous that the IBC could be accepted following a set of short instructions, which are not only plain but also easy to carry out. The IBC requires no special skills, so it is also easy to express and, thereby, to potentially replicate the meme (Heylighen and Chielens 2009: *expressivity*). While talking about simplicity, the briefness of IBC videos should be mentioned as well: The videos usually last between 30 seconds and two minutes and, thus, only demand the attention of potential carriers for a brief period of time. Since attention is a scarce resource, this might be a further advantage over competing memes. Moreover, the IBC meme displays a relatively high degree of proselytism (Heylighen 1999: *proselytism*), as the rules of the IBC include videotaping and publishing the challenge and, especially, nominating three others.

A further criterion for meme replication is that memes are supposed to be more successful if they evoke strong feelings and emotions (Heylighen and Chielens 2009: *affectivity*). The IBC meme stimulates such emotions on various levels. One important supporting element for the IBC meme has been *altruism* (van der Linden 2017): By participating in the challenge and spreading the meme, carriers engaged in a good cause to fight a vicious disease, which seems to have been an important incentive for many people to accept the challenge.[25] However, altruism is certainly not the only factor contributing to the successful spread of memes,[26] but in the case of the IBC meme, an altruistic motivation has arguably been one of the most salient ones, especially for the combination of the IBC with ALS: Never before had ALS gained such immense attention in such a short period of time as during the successful diffusion of the IBC meme during summer 2014. Moreover, there is the empathetic feeling of being poured with ice-cold water and there are myriads of humorous elements that can be put in and transmitted via the videos. Although the unpleasant feeling of pouring ice water over one's head might have barred some people from accepting the challenge, this obstacle may have been mitigated by the IBC meme providing a relatively high degree of (perceived or actual) utility to potential carriers: Among other things, the IBC is an opportunity to commit oneself to a good cause, thereby also giving a good account of oneself, while having some fun (Heylighen 1999: *utility*). There are countless ways for carriers to present themselves "in the proper light" (e.g., as particularly humorous and generous) in their videos, which is closely related to what West (2004) calls "conspicuous compassion" (see also Marsden 2014).

A weakness of the IBC meme is that there are plenty of similar (and also some rival) memes. Distinctiveness constitutes an essential success factor, as distinct memes are

---

[25]On top of that, Blackmore (1999) claims that memes spread by altruistic carriers have a higher probability of being successful than memes spread by non-altruists. According to Blackmore (1999), this is due to altruists being (on average) better connected to other potential carriers because of their caring attitude and their higher chances of being liked by others.

[26]See, for example, Heath et al. (2001), who found disgust to be a central success factor for the spread of urban legends, or Dobele et al. (2007), who examined the role of emotions more generally for passing on viral messages.

supposed to be noticed and understood more easily, which makes their adoption more likely (Heylighen 1999; Spitzberg 2014: *distinctiveness*). Even with the link to ALS in its basic structure, it is not so distinct from the general challenge without its reference to ALS. Indeed, even after the IBC and ALS got linked, not everyone who took up the challenge was aware of that relation or mentioned it in the videos and hashtags (Pressgrove et al. 2018; Rossolatos 2015; Sohn 2017; van der Linden 2017). As already hinted above, the IBC meme cannot be considered as truly new or innovative, since the challenge itself (without its link to ALS) and related memes (e.g., the cold water challenge or the neknomination) have existed before its appearance. This constitutes a further drawback, since a meme is more likely to spark a potential carrier's interest if it is perceived as new (Heylighen 1999: *novelty*). Nevertheless, since the IBC meme received significantly more attention than the cold water challenge or the IBC before fusing with ALS, for most people confronted with the IBC meme, it was actually new. Apparently, the meme found a niche next to existing, rather similar, memes and its success implies that carriers perceived a relative advantage of the IBC meme over the others (Spitzberg 2014: *relative advantage/niche*).

Looking at the *level of the individual (human) carrier* (b), one of the factors that can be assumed to have been most conducive to the IBC meme's successful diffusion is that many celebrities have participated and contributed to its popularity (Burgess et al. 2018). This is in line with findings on (prestige) biases in social learning (Acerbi 2016). As Dawkins and Wong (2016, p. 326) put it: "People are most apt to copy their memes from admired models." Due to the fact that the credibility, reputation, and authority status of a source are essential incentives for potential carriers to accept the IBC or not, without the participation of these prestigious individuals, the population that was reached would arguably have been much smaller (Burgess et al. 2018: *prestige imitation*; Heylighen and Chielens 2009: *authority*; Spitzberg 2014: *source credibility*).

Beyond this, whether a meme is adopted by a potential carrier is significantly influenced by the degree to which a new meme is consistent with the previously acquired knowledge and convictions of a carrier (Heylighen and Chielens 2009: *coherence*; Schlaile et al. 2018: *compatibility*; Spitzberg 2014: *frame resonance*). As already hinted earlier, especially due to sufficient familiarity with prior viral challenge memes and at the same time a sufficiently high degree of novelty, the degree of coherence of the IBC meme can be considered to have been relatively high for most carriers. Naturally, the perceived degree of coherence varies individually due to differences in previously adopted memes, and cultural (in-)coherence could in part explain why the IBC was adopted by a relatively larger population of carriers in some regions, while the interest in the IBC was rather moderate in others (as shown in Fig. 6.4). Moreover, there have also been some rival memes competing with the IBC meme for attention (Spitzberg 2014: *counter-memes and frames*): As it is the nature of memes as informational entities to be in a competition with others for attention (see Sect. 6.2.1), it is no surprise that in the wake of the IBC meme's success there also arose several memes that worked against it. One of these rival memes could be called the "water-saving meme." Water is a scarce resource, in some regions more than in others, so that many people deemed the IBC meme to harm the environment.

Several organizations, therefore, worked against the IBC meme to promote water saving and access to clean water. Another, perhaps less obvious, rival meme that was working against the IBC meme could be the one for preventing animal harm. Among others, People for the Ethical Treatment of Animals (PETA) drew attention to the information that the ALS Association funds experiments on animals, which prompted potentially influential carriers such as Pamela Anderson not to accept the IBC and, in turn, to challenge the ALS Association to stop supporting animal experiments.[27]

Finally, regarding the *network level* (c), it can be stated that the most-involved social media platform Facebook shows a relatively high degree of *homophily* within the (ego-)networks of individual platform users, as these networks consist of actively selected members / "friends." At the same time, Facebook displays a relatively high degree of structural *heterophily* when looking at the entire population of users. Heterophily at the borders of communities facilitates that a meme can cross individual sub-networks and thus diffuse to a greater number of potential carriers (Spitzberg 2014). This combination of homophily and heterophily can be assumed to have constituted a viable basis for the IBC's successful diffusion. Rather trivially, a further advantage of social media platforms such as Twitter or Facebook is their large number of active users (Spitzberg 2014: *number of nodes*). Finally, as many of the early meme adopters were famous individuals, they arguably showed a high centrality and a high potential to influence others (Spitzberg 2014: *number and centrality of influencers*).

6.4 Agent-based Simulation of the IBC Diffusion

In the previous sections, we have shown that many aspects of memetic theory offer useful approaches to explaining the evolution, success, and relative lifespan of viral phenomena. However, a quite substantial shortcoming of the current state of memetic theory is that it does not (yet) provide a generally applicable framework that could be used to formalize the spread of such phenomena either qualitatively or quantitatively. While there exist theoretical frameworks building at least in part on memetic considerations (e.g., Schlaile et al. 2018; Spitzberg 2014) or include evolutionary elements (e.g., Morone and Taylor 2010), such frameworks are usually rather abstract in nature, leaving it unclear how to apply them to real-world phenomena.

Taking the viral nature of the IBC meme seriously, we can observe that the temporal progression of its spread indeed closely follows that of an infectious disease: When recalling the global Google search interest graph presented in Fig. 6.3, it becomes evident that its shape resembles that of the "infected" part in a simple epidemiological SIR[28] model (Kermack and McKendrick 1927; see also Namatame and Chen

---

[27] http://www.peta.org/blog/pamela-anderson-takes-ice-bucket-challenge-animals/. Note that this meme for preventing animal harm is likewise supported by altruism, although in this case the altruistic element is not limited to humans but also includes other species.

[28] SIR stands for susceptible-infected-recovered or susceptible-infected-removed.

2016, Chap. 5) strikingly well. We therefore stand back from our meme-centered perspective for a moment in order to investigate whether the IBC meme's spread can be described as a viral epidemic, or—as its cognitive counterpart is often called in a sociological context—a process of social contagion.

Due to the lack of precise empirical data, we choose here to employ a computational model building on the method of agent-based modeling (ABM). ABM is a powerful technique that has often proven useful for modeling dynamic social and economic phenomena (e.g., Breslin 2014; Goldstone 2005; Mueller and Pyka 2017b; Müller 2017; Schlaile et al. 2018; Wilensky and Rand 2015). While verbal descriptions (such as the ones in Sect. 6.3) can be insightful, one of the major advantages of ABM is that it can be used to design tailored computational experiments allowing to examine the influence of different model parameters on the various aspects of the investigated process (Breslin 2014). According to Chielens and Heylighen (2005), there have also been various agent-based simulations of cultural diffusion processes, where the first one that explicitly mentions memes may probably be Gabora (1995).

In this section, we use the considerations of the previous descriptive memetic analysis in conjunction with the basic rules of the IBC meme to design an agent-based simulation model capable of capturing the temporal progress of the IBC meme's diffusion pattern.[29] Note that, to this aim, it is not necessary to incorporate the evolutionary elements of variation, mutation, or recombination.[30] Nevertheless, several insights from our memetic perspective in the previous section will enter the model's design. In contrast to a merely descriptive analysis, the model will allow us to assess the importance of particularities of the underlying social structure, the influence of individual and aggregate properties of agents, as well as the effect of celebrities. In the following subsection, we begin by describing the basic features of our model, which we then systematically analyze and extend further.

6.4.1 Baseline Scenario

Upon initialization, a single-component (i.e., fully connected) and undirected social network with a selectable number of nodes N (human agents) and links M (connections, e.g., friendships or communication channels between the agents) is created according to one of several selectable options, including but not limited to the common *random*, *small world*, and *preferential attachment* algorithms (for an overview of different network types, see, e.g., Barabási 2016; Newman 2010). For details on the network-generating algorithms we employed, see Appendix D in Schlaile et al. (2018).

As a first setup, we choose a Barabási-Albert (BA) network structure (Barabási and Albert 1999), as its almost scale-free, fat-tailed *degree distribution* is closest to

---

[29]The model is written in NetLogo (Wilensky 1999), version 5.3.1.

[30]As mentioned above, even though considerable variations in the meme's expression were observed, its core elements—including instructions for its spread—remained relatively stable.

that of a real online social network (Grandjean 2016; Mislove et al. 2007; Ugander et al. 2011).

According to Dunbar (1993, 1998, 2009, 2010, 2016), humans have, on average, a stable social network of about 150 people (i.e., *Dunbar's number*). We, therefore, used Dunbar's number to calibrate the number of links between agents so that we arrive at an average degree $\overline{d}$ of 150.[31] Even though a real social network typically comprises millions of users, we restrict the simulated network size to a maximum of $N = 500$ agents due to computational feasibility. Thus, with $\overline{d} = 150$, our first (undirected) network setup consists of $N = 500$ nodes connected by $M = \frac{1}{2}N\overline{d} = 37,500$ links.[32]

One of the central elements of the IBC meme is its chain-letter-like nomination rule.[33] As most variants of the meme contained the conditional rule 'if you just accepted and performed the IBC, then nominate three peers to do the same within 24 hours' (see also Rossolatos 2015), we use this nomination procedure as a reference pattern throughout our simulations.[34]

Initially, the simulation is "seeded" by randomly choosing five agents (henceforth called initiators) who already performed the challenge. Then, at each of the following time steps (in ABM often referred to as ticks), each agent who accepted the challenge at the previous step nominates three of its neighbors to accept the IBC, under the constraint that only agents who have not previously accepted the challenge can be chosen. In Fig. 6.6, which visualizes this procedure, black circles represent agents that have not been nominated, and red circles represent those who accepted the challenge.

The yellow-colored circles depict those agents that have been nominated but did not accept the challenge at that time step. This non-acceptance is due to a property the agents are endowed with in our model, namely the resistance R against the IBC meme. As discussed in the previous section, due to various pre-existing memes, potential carriers may be considered to have some kind of "memetic immune system" that prompts them to resist a meme to a certain degree due to its incoherence or incompatibility with that agent's other memes. Moreover, the utility of participating

---

[31] Note, however, that in the study by Ugander et al. (2011), the average Facebook user was found to have around 190 friends.

[32] We also simulated a network with 2,000 instead of 500 agents (not shown), which does not change the fundamental behavior and dynamics of the diffusion pattern; the system is just scaled and there remain more agents that did not accept the challenge at the end of the simulations. For the sake of computational feasibility, we thus stick to systems with 500 agents.

[33] Remember that Goodenough and Dawkins (1994) already addressed the virality of actual chain letters using the dissemination of the St. Jude chain letter as an example. Although chain letters have existed for quite a long time, the diffusion pattern of chain-letter-like memes can be assumed to be strongly influenced by the medium and the type of network that is used to transfer them, so it is not redundant to address the diffusion patterns of memes exhibiting chain-letter-like instructions again, this time for social media platforms.

[34] Also note that, due to the focus of this paper, at this stage, our model does not simulate donations. These could, however, in principle be modeled in a relatively simple way following the rules of the actual IBC: *if* you get nominated, *then* accept the IBC and randomly donate anything between 1 and 10 units of currency, or *else* donate 100 units of currency.

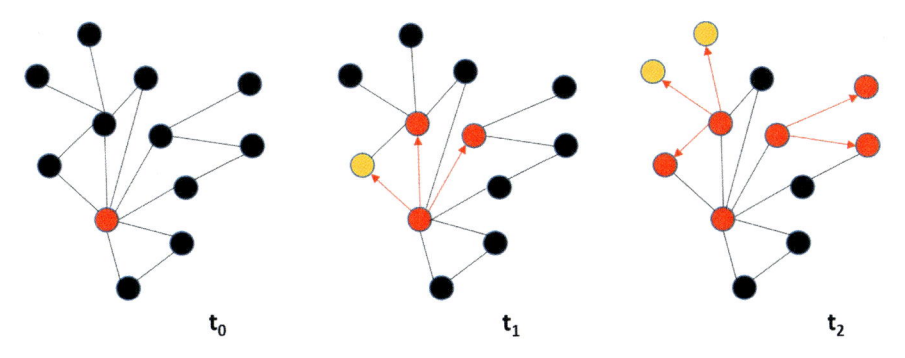

Fig. 6.6 Visualization of the simulated nomination procedure.

in the IBC will be perceived differently by various actors. In our model, the individual value of R is a floating-point number drawn from a uniform random distribution on the interval $[0, 2\overline{R})$, where $\overline{R}$ is an adjustable model parameter denoting the mean resistance.

To capture the nomination procedure while at the same time incorporating a simplified representation of peer pressure or social reinforcement, each time an agent is nominated, its resistance value is reduced by another value E, which we simply call the effect of a nomination. Implicitly, this effect can also be viewed as an aggregate quantity incorporating the supportive characteristics of the IBC meme, including distinctiveness and novelty, simplicity, expressivity, affectivity, etc. If (and only if) an agent's resistance is reduced to zero or lower during nomination, the agent will accept the challenge. For this baseline scenario, we will assume the effect E to be the same for all nominations.

Since the network size is much smaller than that of a real network, we collect statistical averages from one million simulation runs for each investigated parameter set. To be precise, we sample different realizations of the same type of network topology by setting up 100 different network structures and repeating the random initialization and simulation procedure 10,000 times for each realization. Figure 6.7 displays time series of data obtained in this manner as averages over all simulation runs for BA networks with $N = 500$ agents, an average degree of $\overline{d} = 150$, a mean resistance of $\overline{R} = 30$, and a nomination effect of $E = 20$. The number of initiators was set to 5 throughout all our simulations.

The displayed curves are the number of newly performed challenges at any given time step (top left, "newly accepted"), the cumulative number of accepted challenges (top right, "total accepted"), the number of new nominees at each time step (bottom left, "newly nominated"), as well as the total number of nominees existing in the network (bottom right, "total nominated"). Especially the number of newly accepted challenges closely resembles the fast-growing but short-lived diffusion pattern observed in Sect. 6.3.2. From a mathematical point of view, it is the temporal derivative of the total number of accepted challenges, explaining the S-shaped behavior of the latter.

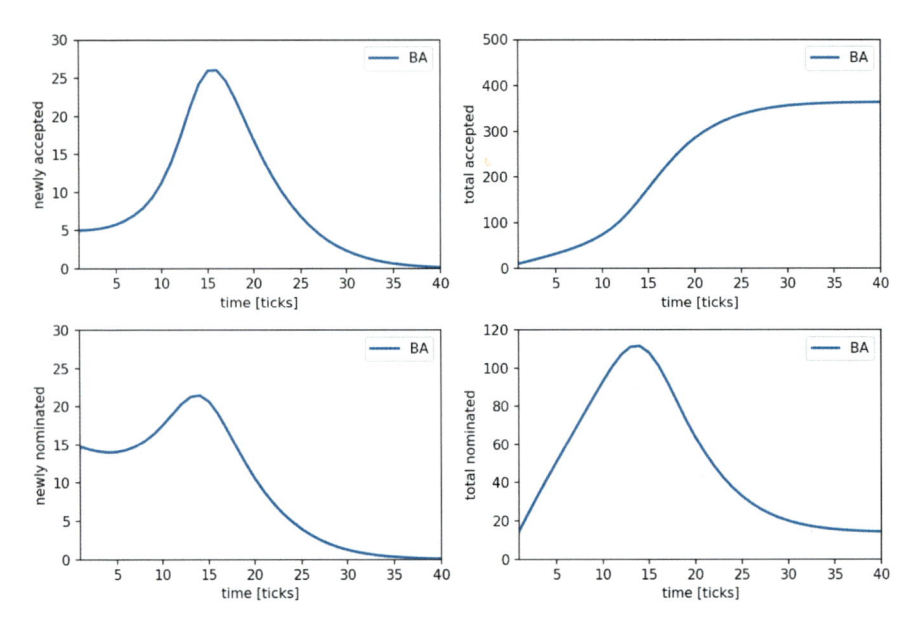

Fig. 6.7 Time series for BA networks. Lines are averages over 100 different networks with 10,000 simulation runs per network (1,000,000 runs in total). The standard error of the mean is smaller than the line width and, therefore, not shown.

It is justified to expect the general interest in the topic to be somehow correlated with the number of newly accepted challenges, since each time someone performs the IBC, this will potentially raise other peoples' attention, affecting at least the ones being nominated thereafter. However, this does not imply that the Google search interest curves shown in Sect. 6.3.2 can be simply interpreted as the temporal derivatives of the real number of people who performed the IBC. In contrast to our model, the real-world interest in the topic is likely to be strongly biased depending on *who* performed the IBC, and mass media reports might generate further interest in the topic from people who could not be reached from within an online social network. The curve depicting the number of newly nominated agents shows a similar but slightly different behavior from the number of newly accepted challenges. This is due to the fact that not all nominated agents accept the IBC right away, and only nominations of agents who have not been nominated before are counted as new. Thus, in the initial stages of the simulations, this number may decrease even though the total number of nominees (i.e., agents who have been nominated but have not accepted the IBC, yet) increases linearly in the beginning. For the same reasons, the number of newly nominated agents does not correspond to the change in the total number of nominees. Nevertheless, the number of new nominations is also expected to have an effect on the general interest in the topic, since newly nominated agents are likely to search the Internet for additional information on the IBC.

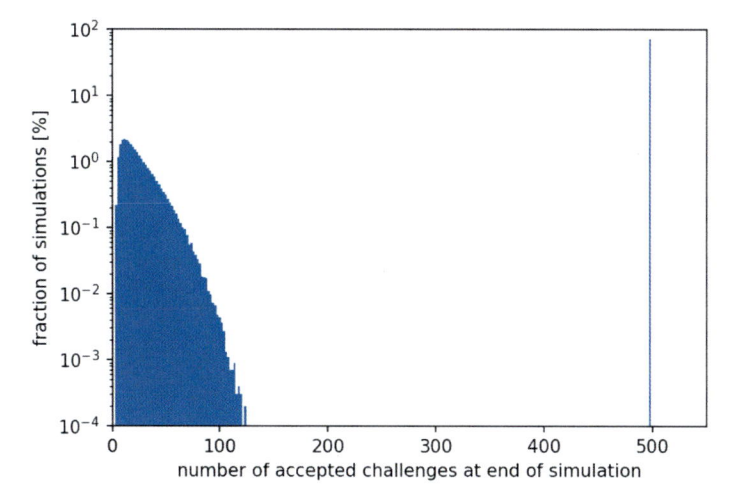

6 Viral Ice Buckets: A Memetic Perspective on the ALS Ice ...

Fig. 6.8 Histogram for the number of carriers who accepted the challenge at the end of the simulation in a BA-type network (note: logarithmic y-axis).

At the end of the simulations, an average of approximately 362 out of 500 agents (72.4%) accepted the IBC, about 13 agents (2.6%) refused to accept despite having been nominated, leaving a fraction of 25% of all agents who have not been reached at all. Since the data shown in Fig. 6.7 are averages over all simulations, they do not provide information on how the success of the IBC, i.e., the final number of accepted challenges, is distributed among individual simulation runs.

The histogram displayed in Fig. 6.8 depicts the relative likelihood (given in percent) of a simulation to end with a certain number of accepted challenges.

In 64% of our simulations, all 500 agents have accepted the challenge, and the remaining 36% of all simulations ended with a number of accepted challenges below 130. In between these numbers, there exists a large gap, indicating that once a threshold or "critical mass" of roughly 130 accepted challenges is reached, all agents in the network will contract the meme. While such a critical mass is likely to exist also for the spread of potentially viral phenomena in the real world, in our simulations, the only reason for the diffusion to stop after having surpassed this tipping point is simply the limited size of the network. To the best of our knowledge, there exists no viral phenomenon that has ever been able to even come close to covering an entire online social network. Thus, there must be other reasons that limit the spread of real viral phenomena including but not limited to cultural distance (remember the Google trends analyses for different countries in Fig. 6.4), generally limited attention of carriers to Internet phenomena, competing memes, and the fact that the IBC meme's central elements may not have varied enough to be considered as sufficiently new,

6.4.2 Influence of Average Resistance

So far, we have shown that our model is capable of qualitatively reproducing the overall diffusion pattern of the IBC, and that there exists a critical number of agents required to accept the challenge in order for the diffusion process not to stop prematurely. In this section, we investigate how the temporal progress of the IBC meme's diffusion is affected by different levels of the resistance $\overline{R}$ to the IBC meme. This model parameter can be regarded as an aggregate quantity representing several not easily quantifiable aspects of human behavior (e.g., including the assumption of a differential fit of the IBC meme with nominees' prior memes), and thus, the choice of a reasonable value is important but not a priori clear.

Figure 6.9 depicts the number of newly accepted challenges over time for different values of the mean resistance $\overline{R}$. As before, each of the curves represents an average over a total of one million simulation runs performed on 100 different BA network topologies, where the number of agents and their average degree as well as the effect of a nomination have been kept constant at $N = 500$, $\overline{d} = 150$, and $E = 20$, respectively.

In the case of no resistance ($\overline{R} = 0$), agents always accept the IBC upon nomination. Consequently, already in the very beginning of the simulation, the number of newly accepted challenges increases rapidly, followed by a sudden drop due to the limited network space. In contrast, very high values of the mean resistance $\overline{R} \geq 34$ prohibit the IBC meme's diffusion almost entirely. Only intermediate values enable our model to reproduce the initially slow "incubation period" of the IBC meme's diffusion as well as its decreasing popularity after having reached its peak. If we assume that one tick in the simulation roughly corresponds to one day in real time, a value of $\overline{R} = 30$ yields a curve representing not only the pattern of the IBC meme's diffusion but also its overall duration comparatively well.[36]

Since the temporal progress of the IBC meme's diffusion is influenced by the mean resistance, this also applies to the resulting cumulative number of accepted challenges at the end of the process. The left panel in Fig. 6.10 displays this number as a function of $\overline{R}$. For $\overline{R} \leq 20$, the challenge is accepted by all agents in the network, followed by a smooth non-linear transition toward a negligible number of agents accepting the IBC for $\overline{R} \geq 40$, further justifying our choice of $\overline{R} = 30$.

---

[35] After all, setting hard bounds to the available network space as an effective way of limiting the IBC meme's spread yields results that are qualitatively comparable to the actual spreading behavior of the IBC.

[36] According to the Google trends and Twitter hashtag analyses as well as the ALS donations discussed in Sect. 6.3.2, the IBC lasted between two and three weeks in the summer of 2014.

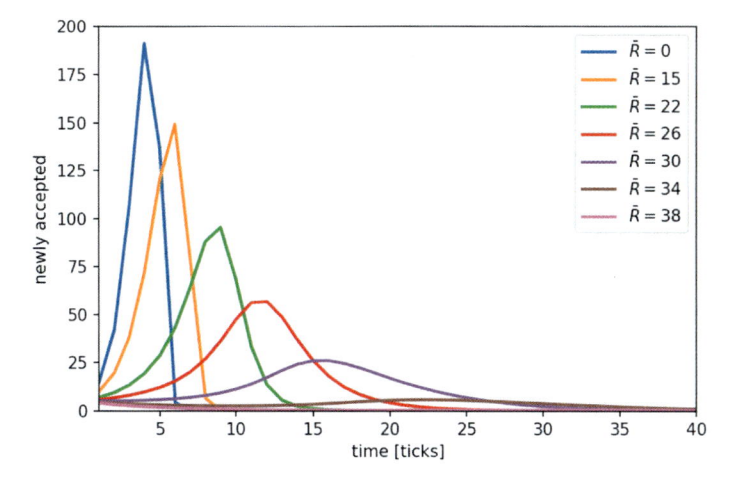

Fig. 6.9 Newly accepted challenges over time for different values of mean resistance in BA-type networks with 500 agents and average degree of 150.

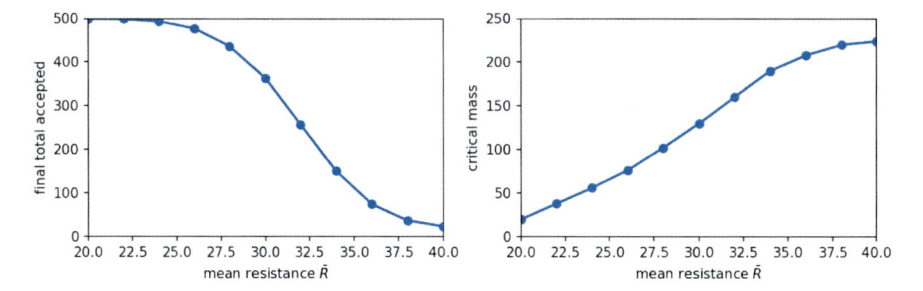

Fig. 6.10 Left panel: Final number of accepted challenges at the end of the simulation as a function of mean resistance; right panel: critical mass needed for the challenge to reach all agents in the network as a function of mean resistance (in a BA-type network).

As we can see in the right panel of Fig. 6.10, also the critical mass of agents required for the challenge to complete is affected by the mean resistance. For low values of the mean resistance ($\overline{R} < 20$), the critical mass is zero and, therefore, this threshold is non-existent. With increasing values of $\overline{R}$, the threshold departs from zero and levels off at a critical mass of 230 agents for $\overline{R} = 40$. For $\overline{R} > 40$, the critical mass cannot be interpreted as such anymore, since none of the simulations was able to surpass this threshold. This means that in order for the meme's diffusion not to die off prematurely, one would essentially need to increase the number of initiators starting the challenge.

6.4.3 Influence of Social Network Topology

As argued in Sect. 6.2.1, memes can be regarded as informational entities. Successful diffusion of information has often been found to depend on the structural characteristics, i.e., the *topology*, of the underlying social network of agents (e.g., Banerjee et al. 2018; Bogner et al. 2018; Mueller et al. 2017; Müller et al. 2014; Watts and Dodds 2007; Weng 2014). This means that although memetics promotes a meme-centered perspective (i.e., the so-called meme's eye view)—and quite justifiably so (Berger and Milkman 2012; Boudry and Hofhuis 2018)—we would get an incomplete picture by disregarding structural properties of the carrier network when analyzing a meme's diffusion (Hui et al. 2018; Schlaile 2018; Schlaile et al. 2018; Spitzberg 2014).

For our analysis of the influence of the social structure, we kept the number of agents and links constant and only varied the social network's topology. More precisely, we used the three common network types of *random graphs* (Erdős and Rényi 1959, 1960), networks with *small-world* characteristics (Watts and Strogatz 1998), and *scale-free networks* (Barabási and Albert 1999), which we denote as ER, WS, and BA, respectively. An overview of the network properties of these three "network archetypes" is given in Table 6.1.

In analogy to Fig. 6.7 in Sect. 6.4.1, Fig. 6.11 shows the time series for 40 ticks based on the averages of 10,000 repetitions on each of 100 networks per network type for the three different network algorithms. As we can see, the qualitative behavior of the diffusion is rather similar (but still statistically different) for all network types.

Table 6.1 Selected properties of the networks

	BA	ER	WS
Number of nodes	500	500	500
Number of links	37,500	37,500	37,500
Density	0.3006	0.3006	0.3006
Average path length	1.6994	1.6994	1.6994
Average clustering coefficient	0.4898	0.3011	0.5214
Average betweenness centrality	174.5	174.5	174.5
Average closeness centrality	0.5941	0.5885	0.5885
Average degree	150	150	150
Median clustering coefficient	0.4931	0.3011	0.5214
Median betweenness centrality	76.50	173.16	173.51
Median closeness centrality	0.5609	0.5884	0.5884
Median degree	108.40	149.91	150.00

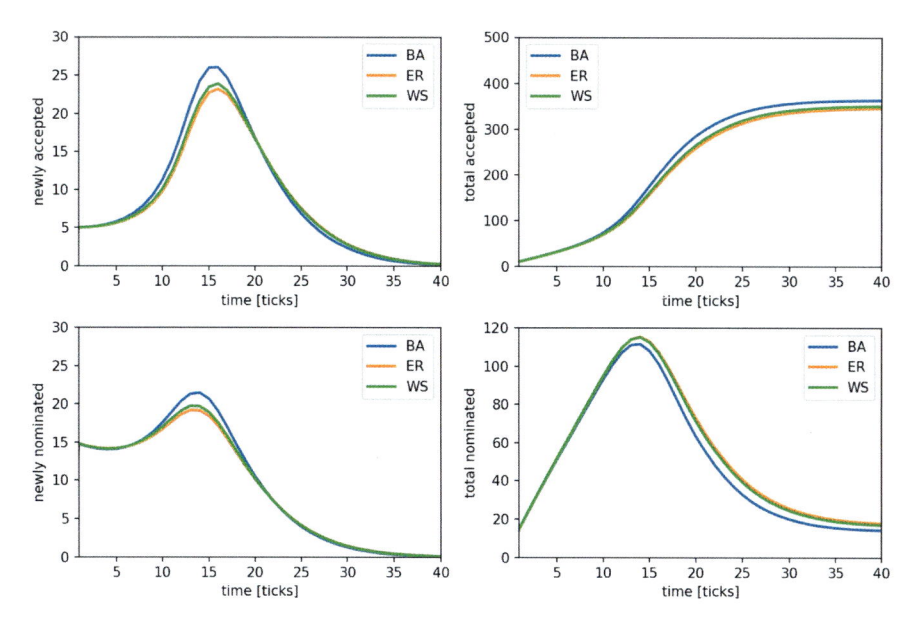

Fig. 6.11 Comparison of diffusion dynamics on three different network types.

As all of the simulated networks exhibit a very high network density, it is particularly interesting that we still observe quantitative differences in the diffusion performance between the three network types, even if these differences are small. In order to discuss these differences and their origins, we will extend our analysis in the following.

Until now, we have kept the average degree $\overline{d}$ constant at a value of 150. Although this number is justifiable based on Dunbar's number, Viswanath et al. (2009, p. 37) argue that "only 30% of Facebook user pairs interact consistently from one month to the next" and that, despite of rapid changes in the number of links, "many graph-theoretic properties of the activity network remain unchanged."[37] Thus, for a more comprehensive topological analysis, we chose to investigate the dependence of the effective diffusion performance on the average degree $\overline{d}$, and, therefore, on network density.[38] To this aim, for each of the investigated network topologies, we performed simulations on networks with constant size but varying average degree, and recorded

---

[37] For comparison, we thus also looked at the diffusion dynamics for networks with an average degree of 40 instead of 150 (not shown). The diffusion dynamics exhibit the same qualitative behavior as with $\overline{d} = 150$, but the magnitude changes slightly: For ER-type networks there are no changes, for BA-type networks there are small changes, while the most notable changes are visible for the WS-type networks. This may indicate that in networks with a lower average degree, the small-world properties could be more important for the IBC meme's diffusion performance than in networks with a higher average degree.

[38] Since we keep the number of agents constant and only vary the average degree, we essentially also vary the network density, which is defined as the ratio of the number of actual links to the number of possible links in the network.

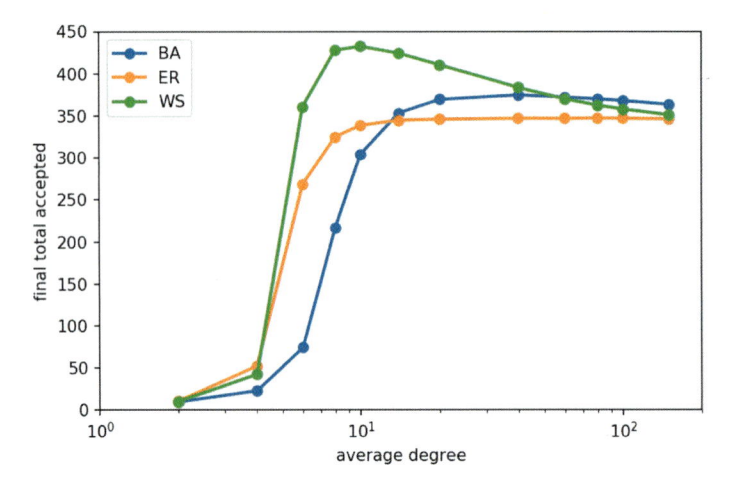

Fig. 6.12 Comparison of total number of accepted challenges at the end of the simulation as a function of average degree for three different network types.

the number of accepted challenges at the end of each simulation as a measure for effective diffusion performance. Figure 6.12 depicts the resulting average final number of accepted challenges as a function of average degree (ranging from 2 to 150) for different network topologies. As before, averages were taken from one million simulation runs per data point.

As it becomes evident from Fig. 6.12, especially in the low-degree region, there exists a general trend for diffusion performance to improve with increasing network density. However, this effect is highly non-linear, with decreasing marginal benefits in the region of high average degrees. Most notably, whereas in both BA and ER networks the final number of accepted challenges follows an almost monotonic S-curve, for WS-type networks, we observe a clearly non-monotonic curve with a distinct maximum at $\overline{d} = 10$ (about 430 or 86 % accepted challenges). These results imply that, independent of network topology, higher network densities are not necessarily beneficial for the meme's diffusion, and diffusion performance is unlikely to solely depend on network density.

Thus, in order to further investigate the topology dependence of diffusion performance, in Fig. 6.13 we show a comparison of different well-established network indices for the three network types as a function of average degree: *average clustering coefficient* (top left), *average closeness centrality* (top right), *average betweenness centrality* (bottom left), and *average path length* (bottom right).

Even though some of these indices are strongly (anti-)correlated, each of them has a distinct meaning. The betweenness centrality of an agent measures the number of shortest paths connecting pairs of other agents in the network which pass through that agent. Thus, agents with a high betweenness centrality can act as *gatekeepers*, i.e., they have a high potential to control the flux of information within the network. In the case of the IBC, this means that such agents may potentially block the spread of

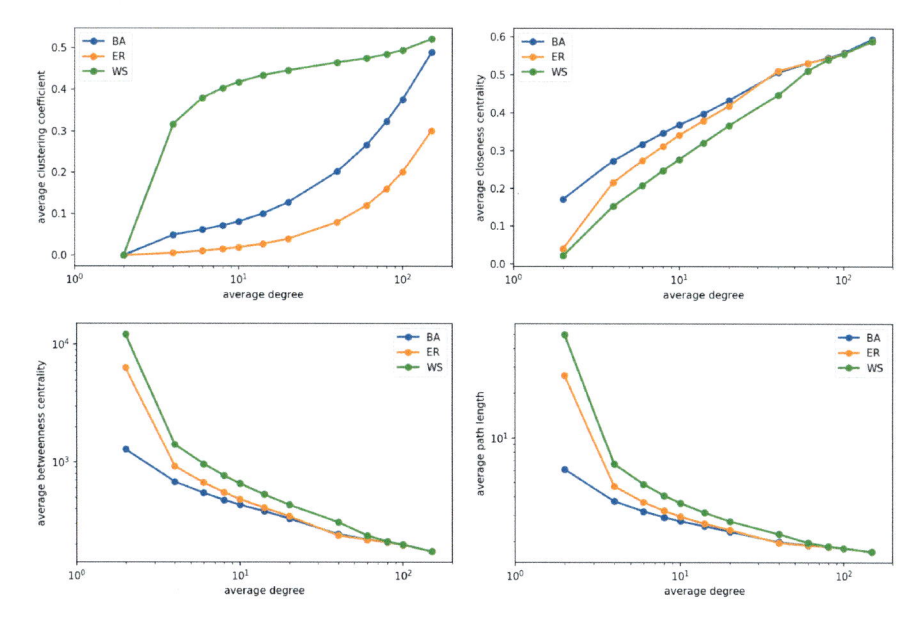

Fig. 6.13 Comparison of average network indices as a function of average degree for three different network types.

the challenge into certain parts of the network, and thus, large average betweenness centralities are disadvantageous for diffusion performance.

The average path length is a global topological observable denoting the average length of shortest paths existing between all pairs of agents in the network, and, therefore, sets limits to the overall speed of information fluxes within the network. However, for the IBC meme's diffusion in our model, the direct impact of this topological index is limited since the speed of diffusion is furthermore bounded by the fact that agents can nominate only up to three neighboring agents.

The closeness centrality of an agent measures the inverse of the average length of shortest paths connecting this agent to the other agents in the network. This means that agents with high closeness centralities have on average shorter connections to other agents in the network and, therefore, such agents can spread information to many parts of the network relatively quickly. Even though one might conjecture that a high average closeness centrality will generally be beneficial for the IBC meme's diffusion performance, we will see later that this is not always the case.

Clearly, the most salient property distinguishing WS-type networks from BA and ER topologies is their average clustering coefficient. An agent's clustering coefficient measures the number of links existing among its neighbors relative to the maximum possible number of such connections. As we have seen, especially in low-density networks, there exist many agents with a high betweenness centrality which might block the IBC meme's spread and thereby increasing its probability to die off prematurely. However, if there exist connections among the neighbors of such agents, these connections provide alternative (at most only marginally longer) paths circumventing the blocking agent. Consequently, a high average clustering coefficient is beneficial for the IBC meme's diffusion performance since the number of such potential gatekeepers is then drastically reduced.

While the much higher average clustering coefficient of WS topologies explains their relative advantage over BA- and ER-type networks in the region of low average degrees, none of the indices displayed in Fig. 13 can directly explain the non-monotonic behavior of the WS curve. In fact, the decreasing diffusion performance of WS-type networks for average degrees $\overline{d} > 10$ is due to a dynamic effect that has been neglected in the discussion until now. As we have seen previously (Sect. 6.4.1), it is vital for the IBC meme's successful diffusion to reach a critical mass of challenge acceptors. This means that its effective performance strongly depends on what happens during the initial stages of the challenge. Since for an average resistance of $\overline{R} = 30$ the individual resistance values R of agents are uniformly distributed on the interval [0, 60), it follows that, on average, only one third of all agents will directly accept the challenge upon first nomination. For another third of the agents, two nominations are required, and the remaining third will only accept the challenge after three nominations. In the very beginning of our simulations, this means that, on average, only one of three nominated agents will directly accept the challenge, but the remaining nominees will have a reduced resistance. In the next step of the challenge, it will then be beneficial if the agents who just accepted the challenge nominate the ones with an already reduced resistance, thereby maximizing the impact of their nominations. This, of course, requires the respective agents to be connected, which further emphasizes the crucial role of a high average clustering coefficient. However, with increasing average degree, it becomes increasingly unlikely for those few agents with an already reduced resistance to be nominated in the early stages of the IBC, thereby making it less likely to reach the required critical mass. Thus, it is the increasing average degree itself that limits the IBC meme's diffusion performance due to the increasingly random choice of new nominees in early stages.[39]

Now that we know that a moderate average degree as well as a high clustering coefficient is beneficial for diffusion performance, we can also explain why BA networks perform slightly better than the other topologies for high average degrees $\overline{d} > 60$. What distinguishes BA from ER and WS topologies at high network densities is the skewness of their almost scale-free (and, therefore, also broader) degree distribution. It follows that in BA networks there exists a particularly large number of agents with a degree below the average, and a relatively small number of agents with a very high degree. This also becomes evident when comparing the values of average with median degree given in Table 6.1. Therefore, a large part of the network suffers less from the previously discussed disadvantage of high degrees. In conjunction with the

---

[39] The resulting non-monotonic behavior of WS-type networks is not visible for BA or ER topologies simply due to the fact that for the latter, diffusion performance is limited by a comparatively low clustering coefficient in the region of low average degrees.

fact that high-density BA networks also have a high average clustering coefficient, this explains their relative advantage over ER or WS topologies for high average degrees.

6.4.4 Influence of Celebrities

The tendency of memetics to focus on the meme's characteristics may also obfuscate the importance of the position or influence of particular carriers. Particularly central agents can serve as influencers or also bottlenecks for a meme's diffusion through social networks. In Spitzberg's (2014, p. 321) words: "The diffusion of memes and knowledge is significantly influenced by individuals who are in a position to control or influence the flow of information throughout a network." In this subsection, we therefore systematically analyze the influence of "celebrities" or prestigious individuals in the network.

Although the literature on influential agents in online social networks mentions various other relevant characteristics, including retweetability, topical similarity, information forwarding activity, and size of cascades (e.g., Hui et al. 2018), probably one of the most important characteristics of such agents is a relatively high degree. Therefore, to capture the difference between prestigious and non-prestigious agents while keeping the model simple, we replace the previously constant effect E of any agent's nomination by a function of its degree d according to

$$E_{\text{celeb}}(d) = E\left(1 + \lambda \frac{d - \bar{d}}{\bar{d}}\right),$$

where $\bar{d}$ denotes the average degree of all agents in the network and λ is an adjustable scaling parameter we set to $\lambda = 0.2$.[40] It should be mentioned that by introducing this degree-dependent celebrity effect, we implicitly included a certain directionality in the network. Even though formally, the network is undirected, it now makes a difference in which direction a connection between agents with different degrees is used to spread the IBC meme.

From the fact that $E_{\text{celeb}}(d)$ is linear in d, and all other parameters entering the equation are constants, it follows that the average effect of all agents in the system is $\overline{E}_{\text{celeb}}(d) = E_{\text{celeb}}(\bar{d}) = E$. This, in turn, means that if the modeled dynamics were not path- and, therefore, topology-dependent, including this effect would yield the exact same results as before. Put differently, if we can observe an influence of the celebrity effect, it must be due to changes in the path- and, thus, topology dependence of the IBC meme's diffusion pattern.

Figure 6.14 compares the temporal progress of the IBC meme's diffusion on different network topologies with and without the celebrity effect for networks of

---

[40]Too high values of λ result in unreasonably fast diffusion for BA networks.

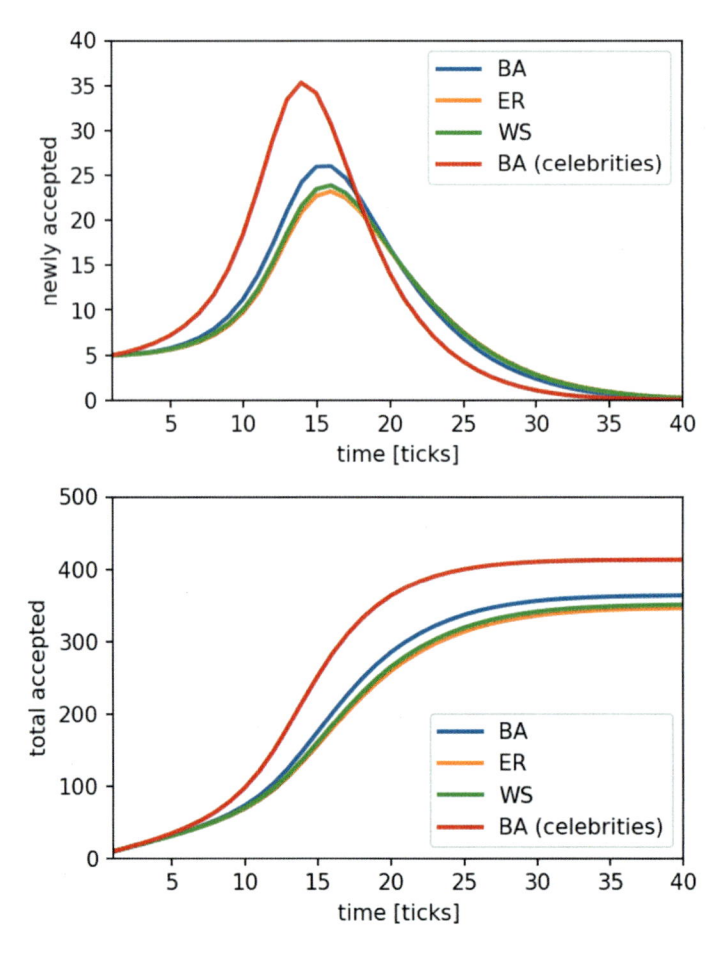

Fig. 6.14 Comparison of diffusion dynamics with and without celebrities (only shown for BA networks as WS and ER networks show no difference with or without celebrities).

500 agents with an average degree of $\bar{d} = 150$. The displayed curves are the number of newly performed challenges at any given time step (top, "newly accepted"), and the cumulative number of accepted challenges (bottom, "total accepted").

For ER and WS networks, the celebrity effect has no influence on diffusion dynamics, which is why only single curves are shown for these network types. The reason why celebrities have no effect for these topologies is due to their relatively narrow and symmetric degree distributions. A narrow degree distribution means that E_{celeb} always remains close to E, and the symmetric distribution of degrees implies that the potentially higher effect of nominations from agents with higher degrees is exactly counterbalanced by agents with lower degree. In contrast, the celebrity effect leads to a faster diffusion and a higher final number of accepted challenges in BA topologies, which is due to the broad and highly asymmetric degree distribution in these networks. Since BA networks are created by a preferential attachment algorithm, these networks are hierarchically structured, leading to an especially high betweenness centrality of agents with high degrees. These agents are, therefore, more likely to be nominated, and once such agents accept the challenge, the further spread of the IBC meme is almost always successful due their high value of E_{celeb}.

6.5 Discussion and Practical Implications

In this section, we reflect on the central points of this work and combine the various findings into a general discussion. At first, we have shown that a memetic perspective is useful for explaining the origin and evolution of the IBC. We have suggested that the IBC meme is a recombination of elements of different prior memes such as the cold water challenge and the neknomination (both of which can be assumed to have evolved from the general idea of a chain letter) in conjunction with the opportunity to stand for a good cause by helping to fight a vicious disease through donations. As discussed in Sect. 6.3, memetic characteristics are also an important part of the explanation of the IBC's success, while the descriptive analysis of Twitter hashtags and Google search interests corroborates these results. Due to its lack of a formalized framework, however, it is impossible to rely exclusively on memetics for investigating the potential influence of the endogenous (meme-centered) characteristics, especially in conjunction with the particularities of the underlying social structure. Arguably, this shortcoming constitutes a substantial point of criticism against a large portion of the memetics literature. Here, we tackle this problem by integrating memetic considerations into the design of an agent-based social contagion model. Admittedly, this approach does not provide a general solution to the missing formalization of the spread of memes as it lacks both a generalizable representation of memes and the evolutionary element of variation (e.g., mutation and recombination). In fact, it is the IBC meme's clear and stable set of instructions that enables us to computationally model its diffusion as a social contagion process.

With our model, we have been able to qualitatively reproduce central elements of the IBC's diffusion pattern. Especially the behavior of the simulated curves for the number of new accepts and the number of new nominees compares well with the actual progression of the IBC as measured by Twitter hashtags, Google search interests, and daily donations.

By systematically analyzing the influence of different model parameters and network types, we gained further important insights: First, the simulation results for our baseline scenario suggest that there are so-called tipping points (e.g., Lamberson and Page 2012), which means that the IBC has to reach a critical mass of carriers in order not to stall prematurely. In our simulations, we employed a fixed network size as an effective measure limiting the IBC meme's spread. In reality, however, in a network space that is larger by many orders of magnitude, there must be other limiting factors. Again, memetic considerations are able to explain the IBC's limited reach and short lifespan. As already mentioned at the end of Sect. 6.4.1, these may include an incoherence with the culture of some countries and counter-memes (which would correspond to a higher resistance in our model) and a decreasing interest of potential carriers in the phenomenon over time due to insufficient variation of the meme in combination with its declining level of novelty.

Furthermore, our simulations have been able to show that networks with a high average clustering coefficient as well as a moderate average degree are beneficial for the IBC meme's diffusion performance.

Moreover, our results suggest that for memes that exhibit a strict nomination rule such as the IBC, hubs mainly have a higher probability of being nominated but they still do not nominate more than three of their peers. Only if we include the assumption that due to their status, hubs also have a higher influence on others (e.g., prestige bias leading to a higher effect of their nominations), we can observe a faster and more wide-ranging diffusion of the IBC in networks exhibiting a highly skewed degree distribution. In fact, it is these characteristics that make online social networks particularly well-suited platforms for the spread of viral phenomena. In this connection, it is also important to mention that the success of the IBC meme was predominantly enabled by the technological advances of modern online media outlets[41] and the increasing interconnectedness among people on the Internet.

Now that we have discussed the advantages and the key findings of our approach, we also have to shed light on its limitations. Probably the most important point of critique one could raise against our model is that it is not "truly" Darwinian and that memetic considerations only find an aggregate representation in the model parameters. Moreover, even when leaving aside the missing evolutionary components of meme diffusion, there exist further aspects that are not properly described by our model. For example, in a more realistic scenario, the distribution of resistance should not be bounded, and its level might potentially be rather similar among cliques of agents due to social homophily. Furthermore, this model parameter should be time-dependent and able to recover if too much time passes between successive nominations. Similar considerations apply to the effect of a nomination, which could decrease over time due to the meme's decreasing level of novelty. Also, in reality, the selection of nominees by agents who completed the challenge is unlikely to happen randomly. This might be particularly important due to the fact that celebrities often nominated other celebrities, which probably increased awareness but did not necessarily entail an imitation by the average Internet user. Moreover, the real spread of the IBC meme has not happened only via the nomination process. Carriers have also adopted the meme after having read articles about the IBC or after having watched videos of prestigious individuals accepting the challenge. These alternative ways of contracting the meme arising from public visibility provided by online social networks have not been incorporated in our simulation model.

Another aspect of the actual diffusion pattern we were unable to reproduce with our model is the second, lower peak that is particularly pronounced in the ALS daily

---

[41] As also Boudry (2018b) notes: "It has been argued that chain letters did not become truly epidemic until the invention of carbon paper. Memes ride piggyback on available technologies" (his footnote 8).

donation data. This second peak can have several explanations. For example, there may have been time lags between nominations and acceptance. The IBC phenomenon may also have received a second wave of attention due to TV reports and other mass media coverage. Another explanation could be that the meme has spread from the United States to other countries which, due to the social network topology and cultural distance, needed a certain time to gain momentum in other countries. Of course, since our simplified model does not incorporate any of these potential factors, we do not observe the second peak in our results.

Drawing on our findings and discussions, we may now compile a number of practical implications for the creation and support of future memetic campaigns promoting a good cause. Note, however, that these implications are far from being exhaustive and should not be considered as some kind of blueprint for successful memes.

(1) *Utilize the self-replicating power of a nomination system:* The IBC meme's chain-letter-like nomination procedure was essential to its rapid diffusion. If every carrier is prompted to spread the meme to several others, propagation accelerates quickly. Particularly Facebook and other social media platforms offer functions to support the progress of such nomination procedures. Note, however, that there may also be an upper bound to an optimal number of nominations since too many nominations may be perceived as too aggressive proselytism or, in other words, spam.

(2) *Add emotional appeal to your meme:* The participation of ALS sufferers (e.g., Pat Quinn or Pete Frates) in the IBC increased the meme's emotional appeal. Memes with a high affectivity are more likely to spread successfully, which means that memes aimed at raising awareness and funds for a good cause should be accompanied by palpable goals and testimonials.

(3) *Engage celebrities:* As the diffusion of the IBC has shown, the heuristic of imitating prestigious individuals has served as a booster for awareness of the meme. However, since celebrities often nominated their peers (i.e., other celebrities), it is important to also engage the general public.

(4) *Keep it simple, stupid (KISS):* As simplicity constitutes an essential success factor for memetic diffusion, potential carriers are more likely to accept memes that carry simple messages. The instructions contained in the IBC meme are short, easy to follow, and the related videos are usually very brief. The difficulty is, therefore, to link the meme to a sufficient amount of accompanying information about a certain kind of charity to raise awareness while at the same time not consuming too much of the potential carriers' attention, which could restrain them from adopting a meme.[42]

(5) *Add an element of novelty and distinctiveness:* New and distinct memes are more likely to attract attention. However, there may be a certain trade-off between newness, distinctiveness, and the compatibility with prior memes (e.g., in terms of coherence). Memes closely related to the IBC such as the *Rice Bucket Challenge*

---

[42] Also note that the abbreviation "tl;dr", which stands for "too long; didn't read," has become very popular on the Internet.

or the *Lather Against Ebola Challenge* mentioned above surely had some success, but their relatively low dissemination may in part be due to having been perceived as mere "copycats" (see also Coscia 2014, on a related note).

(6) *Be prepared for rival memes:* Although the link with an altruistic element may be a factor supporting a meme's diffusion, it does not exclude the meme from competition with other (rival) memes or so-called "counter-memes" (Spitzberg 2014), as was seen in the example of water saving.

(7) *Closely tie the message to the meme:* We have seen that the daily donations appear to correlate more with the course of the #alsicebucketchallenge hashtag than with its counterpart without ALS, and many videos did not mention ALS or donations at all (Sohn 2017; van der Linden 2017). As Sohn (2017, p. S114) puts it: "For some challenge participants, ALS may have been an afterthought, if they thought of it at all." It is, therefore, important to be aware that this detachment can occur and to look for possibilities to intervene and keep the intended effect connected with the meme. Moreover, many Internet memes consist of pictures or videos that are (supposed to be) shared with others. However, although such memes may indeed raise awareness for a given societal problem, they do not directly motivate carriers to take more action than simply spread the meme. For a meme's success in fundraising, it is, therefore, essential to also include an incentive for engaging in the goals of the campaign, which brings us to the next point.

(8) *Incentivize carriers:* Memes that somehow reward their carriers, e.g., by presenting them in a positive way, can be actively exploited for the carriers' tendencies to promote themselves (cf. "conspicuous compassion" Marsden 2014; West 2004).

(9) *Take network topology into account:* As our simulations suggest, network regions with a high average clustering coefficient and a moderate average degree can serve as a fertile soil for the diffusion of the meme, which is of particular importance for the initiation of the campaign.

6.6 Conclusion and Outlook

In this exploratory paper, we have investigated which meme-centered and which structural factors have influenced the IBC's diffusion. We have addressed both questions based on a descriptive as well as a computational pillar. Although we are unaware of any methods for combining both pillars to quantitatively assert which factors have contributed more to the meme's diffusion, we have been able to qualitatively combine the insights from both pillars. While traditional memetic perspectives often tend to (over-)emphasize the characteristics and properties of the meme, diffusion models are likewise often biased toward structural explanations based on properties of the underlying social networks. With our paper, we contribute to this broad literature by combining both perspectives and thereby showing that through synergizing memetic explanations and empirical or computational social contagion research, we gain more insights than the isolated strands of literature can generate on their own.

Although this paper has contributed to studying social learning and cultural evolution by analyzing an actual viral Internet phenomenon in two elaborate ways, there is still much potential for further work. For example, while the agent-based simulation model developed above is suitable for a general analysis of topological effects, future research is needed to better combine the network level with the meme level (e.g., as already theoretically discussed by Spitzberg 2014) in order to also capture the evolutionary dynamics at the levels of the meme and the individual carrier. In this regard, future research endeavors should, for instance, aim at developing methods for quantifying and comparing the criteria for replicative success (including the ones taken up in Sect. 6.3.3). Moreover, our model could be extended in several ways, for instance, by also simulating donations, rival and supporting memes, or by capturing the impact of dynamic resistance and effect values. Yet, even with the present model, a series of additional analyses could be done, for example, to shed light on the actual role of hubs and community structure.

In summary, this paper contributes to the literature on social learning, cultural evolution, and social contagion by providing a solid basis for further research in applied memetics in line with Hull's appeal to memeticists to "shift away from general discussions toward attempts to apply these terms to real cases" (Hull 2000, p. 48).

References

Acerbi, A. (2016). A cultural evolution approach to digital media. *Frontiers in Human Neuroscience*, *10* https://doi.org/10.3389/fnhum.2016.00636.

Adamic, L. A., Lento, T. M., Adar, E., & Ng, P. C. (2016). Information evolution in social networks. In P. N. Bennett, V. Josifovski, J. Neville, & F. Radlinski (Eds.), *Proceedings of the ninth ACM international conference on web search and data mining - WSDM '16* (pp. 473–482). New York: ACM Press.

Álvarez, A. (2004). Memetics: An evolutionary theory of cultural transmission. *Sorites, 15*, 24–28.

Aunger, R. (2002). *The electric meme: A new theory about how we think*. New York: Free Press.

Banerjee, A. V., Chandrasekhar, A. G., Duflo, E., & Jackson, M. O. (2018). Using gossips to spread information: theory and evidence from two randomized controlled trials. *MIT Department of Economics Working Paper*, No. 14–15. https://doi.org/10.2139/ssrn.2425739

Barabási, A.-L. (2016). *Network science*. Cambridge: Cambridge University Press.

Barabási, A.-L., & Albert, R. (1999). Emergence of scaling in random networks. *Science, 286*(5439), 509–512. https://doi.org/10.1126/science.286.5439.509.

Berger, J., & Milkman, K. L. (2012). What makes online content viral? *Journal of Marketing Research, 49*(2), 192–205. https://doi.org/10.1509/jmr.10.0353.

Blackmore, S. (1999). *The meme machine*. Oxford: Oxford University Press.

Blackmore, S. (2000). The meme's eye view. In R. Aunger (Ed.), *Darwinizing culture: The status of memetics as a science* (pp. 25–42). Oxford: Oxford University Press.

Blute, M. (2005). Memetics and evolutionary social science. *Journal of Memetics - Evolutionary Models of Information Transmission, 6*. Retrieved from http://cfpm.org/jom-emit/2005/vol9/blute_m.html

Blute, M. (2010). *Darwinian sociocultural evolution: Solutions to dilemmas in cultural and social theory*. Cambridge: Cambridge University Press.

Bogner, K., Müller, M., & Schlaile, M. P. (2018). Knowledge diffusion in formal networks - The roles of degree distribution and cognitive distance. *International Journal of Computational Economics and Econometrics, 8*(3/4), 388–407. https://doi.org/10.1504/IJCEE.2018.096365.

Boudry, M. (2018a). Invasion of the mind snatchers. On memes and cultural parasites. *Teorema, 37*(2), 111–124.

Boudry, M. (2018b). Replicate after reading: On the extraction and evocation of cultural information. *Biology & Philosophy, 33*, https://doi.org/10.1007/s10539-018-9637-z.

Boudry, M., & Hofhuis, S. (2018). Parasites of the mind. Why cultural theorists need the meme's eye view. *Cognitive Systems Research, 52*, 155–167. https://doi.org/10.1016/j.cogsys.2018.06.010

Breslin, D. (2014). Calm in the storm: Simulating the management of organizational co-evolution. *Futures, 57*, 62–77. https://doi.org/10.1016/j.futures.2014.02.003.

Burgess, A., Miller, V., & Moore, S. (2018). Prestige, performance and social pressure in viral challenge memes: Neknomination, the ice-bucket challenge and smearforsmear as imitative encounters. *Sociology, 52*(5), 1035–1051. https://doi.org/10.1177/0038038516680312.

Cheng, J., Kleinberg, J., Leskovec, J., Liben-Nowell, D., State, B., Subbian, K., & Adamic, L. A. (2018). *Do diffusion protocols govern cascade growth?* Retrieved from http://arxiv.org/pdf/1805.07368v1

Chielens, K. (2003). *The viral aspects of language: A quantitative research of memetic selection criteria* (Masters Thesis, Vrije Universiteit Brussel, Brussels). Retrieved from http://memetics.chielens.net/master/thesis.pdf

Chielens, K., & Heylighen, F. (2005). Operationalization of meme selection criteria: Methodologies to empirically test memetic predictions. In The Society for the Study of Artificial Intelligence and the Simulation of Behaviour (Ed.), *Proceedings of the joint symposium on socially inspired computing* (pp. 14–20). Hatfield: University of Hertfordshire.

Coscia, M. (2013). Competition and success in the meme pool: A case study on quickmeme.com. In *Proceedings of the seventh international AAAI conference on weblogs and social media* (pp. 100–109). Retrieved from https://www.aaai.org/ocs/index.php/ICWSM/ICWSM13/paper/viewFile/5990/6348

Coscia, M. (2014). Average is boring: How similarity kills a meme's success. *Scientific Reports, 4*, 6477. https://doi.org/10.1038/srep06477.

Cronk, L. (1999). *That complex whole: Culture and the evolution of human behavior*. Boulder, CO: Westview Press.

Cullen, B. (1993). The Darwinian resurgence and the cultural virus critique. *Cambridge Archaeological Journal, 3*(2), 179–202. https://doi.org/10.1017/S0959774300000834.

Cullen, B. (2000). *Contagious ideas: On evolution, culture, archaeology, and cultural virus theory. Collected writings* (J. Steele, R. Cullen, & C. Chippindale, Eds.). Oxford: Oxbow Books.

Dawkins, R. (1976). *The selfish gene*. Oxford: Oxford University Press.

Dawkins, R. (1993). Viruses of the mind. In B. Dahlbom (Ed.), *Dennett and his critics: Demystifying mind* (pp. 13–27). Oxford: Blackwell.

Dawkins, R., & Wong, Y. (2016). *The ancestor's tale: A pilgrimage to the dawn of life* (2nd ed.). London: Weidenfeld & Nicolson.

Dennett, D. C. (1995). *Darwin's dangerous idea: Evolution and the meanings of life*. New York: Simon & Schuster.

Dennett, D. C. (2003). *Freedom evolves*. London: Penguin.

Dennett, D. C. (2017). *From bacteria to Bach and back: The evolution of minds*. New York: W.W. Norton & Company.

Distin, K. (2005). *The selfish meme: A critical reassessment*. Cambridge: Cambridge University Press.
Distin, K. (2014). Foreword to the chinese translation of The Selfish Meme. Cambridge University Press and Beijing World Publishing. Retrieved from https://www.distin.co.uk/kate/pdf/Foreword_Chinese.pdf
Dobele, A., Lindgreen, A., Beverland, M., Vanhamme, J., & van Wijk, R. (2007). Why pass on viral messages? because they connect emotionally. *Business Horizons, 50*(4), 291–304. https://doi.org/10.1016/j.bushor.2007.01.004.
Dobson, S., & Sukumar, A. (2018). Memes and civic action: Building and sustaining civic empowerment through the internet. In C. Yamu, A. Poplin, O. Devisch, & G. de Roo (Eds.), *The virtual and the real in planning and urban design: Perspectives, practices and applications* (pp. 267–278). London: Routledge.
Dunbar, R. I. M. (1993). Coevolution of neocortical size, group size and language in humans. *Behavioral and Brain Sciences, 16*(4), 681. https://doi.org/10.1017/S0140525X00032325.
Dunbar, R. I. M. (1998). The social brain hypothesis. *Evolutionary Anthropology, 6*(5), 178–190.
Dunbar, R. I. M. (2009). The social brain hypothesis and its implications for social evolution. *Annals of Human Biology, 36*(5), 562–572. https://doi.org/10.1080/03014460902960289.
Dunbar, R. I. M. (2010). *How many friends does one person need? Dunbar's number and other evolutionary quirks*. Cambridge: Harvard University Press.
Dunbar, R. I. M. (2016). Do online social media cut through the constraints that limit the size of offline social networks? *Royal Society Open Science, 3*(1). https://doi.org/10.1098/rsos.150292
Edmonds, B. (2002). Three challenges for the survival of memetics. *Journal of Memetics - Evolutionary Models of Information Transmission, 6*. Retrieved from http://cfpm.org/jom-emit/2002/vol6/edmonds_b_letter.html
Edmonds, B. (2005). The revealed poverty of the gene-meme analogy - why memetics per se has failed to produce substantive results. *Journal of Memetics - Evolutionary Models of Information Transmission, 9*. Retrieved from http://cfpm.org/jom-emit/2005/vol9/edmonds_b.html
Erdős, P., & Rényi, A. (1959). On random graphs I. *Publicationes Mathematicae, 6*, 290–297. Retrieved from https://users.renyi.hu/~p_erdos/1959-11.pdf
Erdős, P., & Rényi, A. (1960). On the evolution of random graphs. *A Matematikai Kutató Intézet Közleményei, 5*, 17–61. Retrieved from https://users.renyi.hu/~p_erdos/1960-10.pdf
Fawzy, F. (2016). *Ice bucket challenge's 2nd anniversary celebrates its gene discovery*. CNN. Retrieved from https://edition.cnn.com/2016/07/27/health/als-ice-bucket-challenge-funds-breakthrough/index.html.
Fullman, A. (2015). *That's pretty cool: A look at the ALS ice bucket challenge as a rhetorical moment*. M.A. Thesis, University of Nebraska at Omaha.
Gabora, L. (1995). Meme and variations: A computational model of cultural evolution. In L. Nadel & D. L. Stein (Eds.), *1993 lectures in complex systems* (pp. 471–485). Reading, MA: Addison-Wesley.
Gabora, L. (2008). The cultural evolution of socially situated cognition. *Cognitive Systems Research, 9*(1–2), 104–114. https://doi.org/10.1016/j.cogsys.2007.05.004.
Gladwell, M. (2000). *The tipping point: How little things can make a big difference*. New York: Little, Brown and Company.
Goldstone, R. L., & Janssen, M. A. (2005). Computational models of collective behavior. *Trends in Cognitive Sciences, 9*(9), 424–430. https://doi.org/10.1016/j.tics.2005.07.009.
Goodenough, O. R., & Dawkins, R. (1994). The 'St Jude' mind virus. *Nature, 371*(6492), 23–24. https://doi.org/10.1038/371023a0.
Grandjean, M. (2016). A social network analysis of Twitter: Mapping the digital humanities community. *Cogent Arts & Humanities, 3*(1), 361. https://doi.org/10.1080/23311983.2016.1171458.
Grimmelmann, J. (2018). The platform is the message. *Georgetown Law Technology Review, 2* (2), 217–233. Retrieved from https://www.georgetownlawtechreview.org/wp-content/uploads/2018/07/2.2-Grimmelmann-pp-217-33.pdf

Heath, C., Bell, C., & Sternberg, E. (2001). Emotional selection in memes: The case of urban legends. *Journal of Personality and Social Psychology, 81*(6), 1028–1041. https://doi.org/10.1037//0022-3514.81.6.1028.

Heath, J. (2015). Methodological individualism. In E. N. Zalta (Ed.), *The Stanford encyclopedia of philosophy*, spring 2015 ed. Metaphysics Research Lab, Stanford University. Retrieved from https://plato.stanford.edu/archives/spr2015/entries/methodological-individualism/

Heylighen, F. (1993). Selection criteria for the evolution of knowledge. In *Proceedings of the 13th international congress on cybernetics* (pp. 524–528). Namur: Association Internat. de Cybernétique.

Heylighen, F. (1997). Objective, subjective, and intersubjective selectors of knowledge. *Evolution and Cognition, 3*(1), 63–67.

Heylighen, F. (1999). What makes a meme successful? selection criteria for cultural evolution. In *Proceedings of the 15th international congress on cybernetics* (pp. 418–423). Namur: Association Internat. de Cybernétique.

Heylighen, F., & Chielens, K. (2009). Evolution of culture, memetics. In R. A. Meyers (Ed.), *Encyclopedia of complexity and systems science* (pp. 3205–3220). New York: Springer.

Hodgson, G. M., & Knudsen, T. (2010). *Darwin's conjecture: The search for general principles of social and economic evolution*. Chicago: University of Chicago Press.

Hofstadter, D. R. (1985). *Metamagical themas: Questing for the essence of mind and pattern*. New York: Basic Books.

Hrastelj, J., & Robertson, N. P. (2016). Ice bucket challenge bears fruit for amyotrophic lateral sclerosis. *Journal of Neurology, 263*(11), 2355–2357. https://doi.org/10.1007/s00415-016-8297-7.

Hui, P.-M., Weng, L., Sahami Shirazi, A., Ahn, Y.-Y., & Menczer, F. (2018). Scalable detection of viral memes from diffusion patterns. In S. Lehmann & Y.-Y. Ahn (Eds.), *Complex spreading phenomena in social systems* (pp. 197–211). Cham: Springer.

Hull, D. L. (2000). Taking memetics seriously: Memetics will be what we make it. In R. Aunger (Ed.), *Darwinizing culture: The status of memetics as a science* (pp. 43–67). Oxford: Oxford University Press.

Jang, S. M., Park, Y. J., & Lee, H. (2017). Round-trip agenda setting: Tracking the intermedia process over time in the ice bucket challenge. *Journalism: Theory, Practice & Criticism, 18*(10), 1292–1308. https://doi.org/10.1177/1464884916665405

Kermack, W. O., & McKendrick, A. G. (1927). A contribution to the mathematical theory of epidemics. *Proceedings of the Royal Society A: Mathematical, Physical and Engineering Sciences, 115*(772), 700–721. https://doi.org/10.1098/rspa.1927.0118.

Kirby, J., & Marsden, P. (Eds.). (2006). *Connected marketing: The viral, buzz and word of mouth revolution*. Amsterdam: Butterworth-Heinemann.

Lamberson, P. J., & Page, S. E. (2012). Tipping points. *Quarterly Journal of Political Science, 7*(2), 175–208. https://doi.org/10.1561/100.00011061.

Levin, J. (2014). *Who invented the Ice Bucket Challenge? A search for the fundraising phenomenon's cold, soaked patient zero*. SLATE. Retrieved from http://www.slate.com/articles/technology/technology/2014/08/who_invented_the_ice_bucket_challenge_a_slate_investigation.html.

Lynch, A. (1996). *Thought contagion: How belief spreads through society. The new science of memes*. New York: Basic Books.

Mahoney, J. (2015). *The ice bucket challenge and its name calling strategy through social media: A descriptive statistical analysis* (M.S. Thesis, University of Louisiana at Lafayette). Retrieved from https://pqdtopen.proquest.com/doc/1708664845.html?FMT=ABS

Marsden, P. (1998). Memetics and social contagion: two sides of the same coin? *Journal of Memetics - Evolutionary Models of Information Transmission, 2*. http://cfpm.org/jom-emit/1998/vol2/marsden_p.html

Marsden, P. (1999). Book review of Aaron Lynch: Thought contagion: How belief spreads through society. *Journal of Artificial Societies and Social Simulation, 2*(2). Retrieved from: http://jasss.soc.surrey.ac.uk/2/2/review4.html

Marsden, P. (2014). *The future of digital content in sports sponsorship*. Deutsche Sponsoringtage 2014. Frankfurt am Main. Retrieved from: https://de.slideshare.net/Syzygy_Group/the-future-of-digital-content-in-sports-sponsorship

Milner, R. M. (2016). *The world made meme: Public conversations and participatory media*. Cambridge: MIT Press.

Mislove, A., Marcon, M., Gummadi, K.P., Druschel, P., & Bhattacharjee, B. (2007). Measurement and analysis of online social networks. In C. Dovrolis & M. Roughan (Eds.), *Proceedings of the 7th ACM SIGCOMM conference on Internet measurement - IMC '07* (pp. 29–42). New York: ACM Press.

Mitchell, P. (2012). *Contagious metaphor*. London: Bloomsbury Publishing.

Morone, P., & Taylor, R. (2010). *Knowledge diffusion and innovation: Modelling complex entrepreneurial behaviours*. Cheltenham: Edward Elgar.

Mueller, M., Bogner, K., Buchmann, T., & Kudic, M. (2017). The effect of structural disparities on knowledge diffusion in networks: An agent-based simulation model. *Journal of Economic Interaction and Coordination*, *12*(3), 613–634. https://doi.org/10.1007/s11403-016-0178-8.

Mueller, M., & Pyka, A. (2017). Economic behaviour and agent-based modelling. In R. S. Frantz, S.-H. Chen, K. Dopfer, F. Heukelom, & S. Mousavi (Eds.), *Routledge handbook of behavioral economics* (pp. 405–415). Routledge international handbooks.

Müller, M. (2017). *An agent-based model of heterogeneous demand*. Wiesbaden: Springer.

Müller, M., Buchmann, T., & Kudic, M. (2014). Micro strategies and macro patterns in the evolution of innovation networks: An agent-based simulation approach. In N. Gilbert, P. Ahrweiler, & A. Pyka (Eds.), *Simulating knowledge dynamics in innovation networks* (pp. 73–95). Berlin: Springer.

Namatame, A., & Chen, S.-H. (2016). *Agent-based modeling and network dynamics*. Oxford: Oxford University Press.

Newman, M. E. J. (2010). *Networks: An introduction*. Oxford: Oxford University Press.

Pressgrove, G., McKeever, B. W., & Jang, S. M. (2018). What is contagious? Exploring why content goes viral on Twitter: A case study of the ALS Ice Bucket Challenge. *International Journal of Nonprofit and Voluntary Sector Marketing*, *23*(1), e1586. https://doi.org/10.1002/nvsm.1586.

Rossolatos, G. (2015). The ice-bucket challenge: The legitimacy of the memetic mode of cultural reproduction is the message. *Signs and Society*, *3*(1), 132–152. https://doi.org/10.1086/679520.

Røvik, K. A. (2011). From fashion to virus: An alternative theory of organizations' handling of management ideas. *Organization Studies*, *32*(5), 631–653. https://doi.org/10.1177/0170840611405426.

Sampson, T. D. (2012). *Virality: Contagion theory in the age of networks*. Minneapolis: University of Minnesota Press.

Schlaile, M.P. (2018). *A case for (econo-)memetics: Why we should not throw the baby out with the bathwater*. [Conference presentation] The Generalized Theory of Evolution Conference, Duesseldorf Center for Logic and Philosophy of Science (DCLPS), Duesseldorf, Germany.

Schlaile, M. P., & Constantinescu, L. (2016). Exploring the potential of organizational memetics: A review and case example. *Academy of Management Proceedings*, *2016*, 17407. https://doi.org/10.5465/ambpp.2016.17407abstract.

Schlaile, M. P., & Ehrenberger, M. (2016). Complexity, cultural evolution, and the discovery and creation of (social) entrepreneurial opportunities: Exploring a memetic approach. In E. S. C. Berger & A. Kuckertz (Eds.), *Complexity in entrepreneurship, innovation and technology research* (pp. 63–92). Cham: Springer.

Schlaile, M. P., Zeman, J., & Mueller, M. (2018). It's a match! Simulating compatibility-based learning in a network of networks. *Journal of Evolutionary Economics*, *28*(5), 1111–1150. https://doi.org/10.1007/s00191-018-0579-z.

Secretan, J. (2013). Stigmergic dimensions of online creative interaction. *Cognitive Systems Research*, *21*, 65–74. https://doi.org/10.1016/j.cogsys.2012.06.006.

Segev, E., Nissenbaum, A., Stolero, N., & Shifman, L. (2015). Families and networks of internet memes: The relationship between cohesiveness, uniqueness, and quiddity concreteness. *Journal of Computer-Mediated Communication, 20*(4), 417–433. https://doi.org/10.1111/jcc4.12120.

Shifman, L. (2014). *Memes in digital culture*. Cambridge, MA: MIT Press.

Sohn, E. (2017). Fundraising: The ice bucket challenge delivers. *Nature, 550*(7676), S113–S114. https://doi.org/10.1038/550S113a.

Speel, H.-C. (1999). Memetics: On a conceptual framework for cultural evolution. In F. Heylighen, J. Bollen, & A. Riegler (Eds.), *The evolution of complexity* (pp. 229–254). Dordrecht: Kluwer.

Spitzberg, B. H. (2014). Toward a model of meme diffusion (M^3D). *Communication Theory, 24*(3), 311–339. https://doi.org/10.1111/comt.12042.

Ugander, J., Karrer, B., Backstrom, L., & Marlow, C. (2011). The anatomy of the Facebook social graph. Retrieved from: http://arxiv.org/pdf/1111.4503v1

van der Linden, S. (2017). The nature of viral altruism and how to make it stick. *Nature Human Behaviour, 1*(3). https://doi.org/10.1038/s41562-016-0041

Velikovsky, J. T. (2016). The holon/parton theory of the unit of culture (or the meme, and narreme). In A. M. Connor & S. Marks (Eds.), *Creative technologies for multidisciplinary applications* (pp. 208–246). https://doi.org/10.4018/978-1-5225-0016-2.ch009

Velikovsky, J. T. (2018). The holon/parton structure of the meme, or the unit of culture. In M. Khosrow-Pour (Ed.), *Encyclopedia of information science and technology* (pp. 4666–4678). https://doi.org/10.4018/978-1-5225-2255-3.ch405

Viswanath, B., Mislove, A., Cha, M., & Gummadi, K.P. (2009). On the evolution of user interaction in Facebook. In J. Crowcroft & B. Krishnamurthy (Eds.), *Proceedings of the 2nd ACM workshop on online social networks - WOSN '09* (pp. 37–42). New York: ACM Press.

Watts, D. J., & Dodds, P. S. (2007). Influentials, networks, and public opinion formation. *Journal of Consumer Research, 34*(4), 441–458. https://doi.org/10.1086/518527.

Watts, D. J., & Strogatz, S. H. (1998). Collective dynamics of 'small-world' networks. *Nature, 393*(6684), 440–442. https://doi.org/10.1038/30918.

Weng, L. (2014). *Information diffusion on online social networks* (Doctoral dissertation, Indiana University). http://lilianweng.github.io/papers/weng-thesis-single.pdf

Weng, L., Flammini, A., Vespignani, A., & Menczer, F. (2012). Competition among memes in a world with limited attention. *Scientific Reports, 2*(1), 57. https://doi.org/10.1038/srep00335.

West, P. (2004). *Conspicuous compassion: Why sometimes it really is cruel to be kind*. London: Civitas, Institute for the Study of Civil Society.

Wicks, P. (2014). The ALS ice bucket challenge - can a splash of water reinvigorate a field? *Amyotrophic Lateral Sclerosis and Frontotemporal Degeneration, 15*(7–8), 479–480. https://doi.org/10.3109/21678421.2014.984725.

Wiggins, B. E., & Bowers, G. B. (2015). Memes as genre: A structurational analysis of the memescape. *New Media & Society, 17*(11), 1886–1906. https://doi.org/10.1177/1461444814535194.

Wilensky, U. (1999). *NetLogo:* http://ccl.northwestern.edu/netlogo. Evanston, IL: Center for Connected Learning and Computer-Based Modeling, Northwestern University.

Wilensky, U., & Rand, W. (2015). *An introduction to agent-based modeling: Modeling natural, social, and engineered complex systems with NetLogo*. Cambridge: MIT Press.

Chapter 7
General Discussion: Economemetics and Agency, Creativity, and Normativity

Michael P. Schlaile

Abstract This chapter addresses some important issues that have not received sufficient attention in the previous chapters. It should be noted, however, that each topic would justify a full paper or perhaps even a book in its own right. Therefore, this general discussion strives for conciseness rather than completeness. Section 7.1 sets off with arguably one of the most widely debated issues in social science, namely the question of agency versus structure—yet, in this case, with a particular focus on the agency of memes. Section 7.2 addresses the implications of a memetic perspective on creativity and novelty. In Sect. 7.3, we turn to some of the normative implications of the (econo-)memetic approach.

7.1 Agency—Memes as Quasi-Agents?

As already mentioned above, the discussion about whether human agency (usually associated with notions such as freedom of choice and of will, intentionality, indeterminism, etc.) or the social structure (e.g., institutions, culture, social networks, path dependence, etc.) has more influence on the behavior or actions of individuals is nothing new (e.g., Archer 1996; Blute 2010, Chap. 6; Callinicos 2004; Hodgson 2004b; see also Barker 2012, pp. 237–251 for a brief introduction). What is new in the light of memetics, however, is that now structural or cultural elements—memes—are frequently depicted as agents that intentionally "infect" human minds in the interest of getting copied. Or, as Robert Aunger boldly summarizes this perspective:

> In short, we don't have ideas; ideas have us! We are hosts to parasites feeding on our brains that cause us to behave in ways beneficial to them, not us (Aunger 2002, p. 18).

M. P. Schlaile (✉)
Department of Innovation Economics, University of Hohenheim, Stuttgart, Germany
e-mail: schlaile@uni-hohenheim.de

© The Author(s), under exclusive license to Springer Nature Switzerland AG 2021
M. P. Schlaile (ed.), *Memetics and Evolutionary Economics*, Economic Complexity and Evolution, https://doi.org/10.1007/978-3-030-59955-3_7

Alternatively, we may turn to Daniel Dennett's only slightly less disturbing slogan: "A scholar is just a library's way of making another library" (Dennett 1995, p. 346). This view can be seen as one extreme end of the spectrum—the "intentional stance" (Dennett 1987) gone rogue, if you will. However, we may also remember from Chaps. 1 and 2 that regarding culture as an independent and even harmful entity is not an invention of memeticists. One might, therefore, argue that the agency problem is not an inherently memetic issue.

At the other end of the spectrum are the cultural instrumentalists who emphasize the purposeful and conscious design and transmission of cultural ideas and artifacts. For example, as Tim Ingold claims:

> Far from being a passive medium, the mind—or consciousness—is an active agent, engaging its powers of reason in the selection of cultural objects and operating over the long course of history through the *instrumentality* of such objects (Ingold 1986, p. 66, italics in original).

In the same vein, John Searle (1997) argues with regard to memes that the "spread of ideas and theories by 'imitation' is typically a conscious process directed toward a goal" (p. 105). This mindset is also echoed by Douglass North (2005) when he claims that "institutions must be explained in terms of the intentionality of humans" (p. 42).[1]

Some of the early debates in the *Journal of Memetics* about a memetic concept of "self" point to a similar disagreement over the role of agency (e.g., Hull 1999; Price 1999; Rose 1998, and others). Obviously, we cannot resolve these age-old debates in this short section. I would, however, argue that—as often—the "truth" probably lies somewhere in the middle. At this juncture, it may thus be important to revisit Blackmore's quote we already encountered in Sect. 3.2:

> [N]o meme theorist is likely to reject human goals as irrelevant to memetic evolution. The interesting question is what role they play. Are human goals the ultimate design force for culture ... or are they just one of many factors in a Darwinian design process acting on memes (Blackmore 2005, p. 409)?

In terms of the more general agency/structure debate, for example, Chris Barker makes a related point:

> Those persons whose acculturation has led them to be highly educated in a formal sense, or who have accrued wealth, may have more options for action than others. The idea of agency as 'could have acted differently' avoids some of the problems of 'free as undetermined' because the pathways of action are themselves socially constituted (Barker 2012, p. 241).

Of course, very quickly the issue of agency can then also become one of power, but we shall not get lost in conceptual nuances, here.[2] As I have already proposed elsewhere

---

[1] However, for a critical account of North's approach to cultural evolution, see Krul (2018, Chap. 5); and also remember Rosenberg's (2017) objection: "The grip of human intentionality on the explanatory strategies of the social sciences has always been strong, and has strengthened in recent decades owing to the intellectual imperialism of economics, and the attendant prestige of rational choice theory" (p. 347).
[2] Readers interested in the issue of power in the context of agency may, for example, turn to Archer (1996), Dietz and Burns (1992), or Herrmann-Pillath (2013, 2018), for more insights.

(Schlaile and Ehrenberger 2016), (i-)memes (in the sense of the mental representations of a meme) can become part of schemata that influence both the search space for new information and behavioral heuristics, thereby affecting the possibilities or opportunity space of an economic actor (e.g., an entrepreneur).[3] Consequently, memes as elements of cultural schemata have some form of proto-agency by virtue of their causal influence on the "could have acted differently" part in Barker's above sense.[4]

An important and helpful discussion on agency in our context can also be found in the works of Tom Burns and Thomas Dietz (Burns and Dietz 1992; Dietz and Burns 1992), especially since their Darwinian approach and their understanding of rules allows us to link back to memes and the rule-based approach to evolutionary economics (as discussed in Chap. 3).[5] Although Burns and Dietz are not concerned with the agency of rules themselves, their arguments for the limitations of human agency due to the differential evolution and structuration of rules help us to realize that human agency is highly situation-dependent. Dietz and Burns (1992) offer four criteria for agency:

> Agency requires that actions be *effective* in changing material or cultural conditions, that they be *intentional*, sufficiently *unconstrained* that actions are not perfectly predictable and that the actor possesses the ability to observe the consequences of an action and be *reflexive* in evaluating them (Dietz and Burns 1992, p. 194, italics in original).

If we accept these criteria, it will quickly become clear that memes can only fulfill some of them, and others only to some degree. It may sound like a bold claim to hard-core meme enthusiasts, but a meme cannot *seriously* be considered as really, intrinsically intentional or reflexive. However, that is *not* to say that the "intentional stance" or the meme's eye view is useless (e.g., see Boudry and Hofhuis 2018; Hofhuis and Boudry 2019, for an illuminating case for the meme's eye view), but to simply state that memes do not satisfy the four criteria for agency proposed by Dietz and Burns (1992).

It should also be noted that for Burns and Dietz (1992, p. 273) agency constitutes "a continuous rather than a categorical property of actors." Interestingly, Stephan Fuchs (2001) takes this argument even further and maintains:

> Agency, creativity, and genius are not essential properties that some persons 'have' qua person. Rather, they are attributions and dependent variables, more likely in some situations, on some occasions, and in some networks than others. ... Agency and structure, and micro/macro, are ... variations along a continuum (Fuchs 2001, p. 39, emphasis removed).

---

[3] Think of a simple example: The members of an isolated indigenous tribe have no concept of airplanes. It is not very far-fetched to argue that the probability for the tribe members to develop any desire to engage in air travel approaches zero—unless they "catch" the relevant mental representation(s) "enabling" them to think of flying with an airplane. As Ball (1984, p. 159) puts it: "We have a good deal of freedom to do what we want but much less freedom to want what we want."

[4] Or, if you want to take up the terminology Steven Johnson borrowed from Stuart Kauffman, these cultural schemata affect "the adjacent possible" (Johnson 2010, Chap. 1; Kauffman 2000).

[5] In fact, one may substitute the word rule with meme in the two papers by Burns and Dietz and still have a perfectly plausible argument.

How can the issue of agency of memes be resolved, then? It is quite obvious that they cannot be regarded as actors with the same type of agency as humans (remember also the sometimes fuzzy distinctions between replicators and interactors mentioned in Chap. 2). However, they can still be causally responsible for (changes in) human behavior as instructions or rules, which is why I would call memes "quasi-agents".

There may be (at least) two promising ways of conceptualizing memetic "agency" that could be explored in future research. The first one could draw on the notion of *affordance* (Gibson 2015; McGrenere and Ho 2000; Nye 2011) in the sense that memes *afford* actions and behaviors.[6] Memes as affordances do not imply any volition, intentionality, or consciousness on the part of the memes; yet, they can still be regarded as independent entities influencing the perceptions and action possibilities of a human actor.

The second way could be to employ *actor-network theory* (Latour 2005), which is fairly well known from science and technology studies (see also Bijker and Law 1992; Blok et al. 2020). The particular concept I have in mind for the meme is that of an *actant*. According to Madeleine Akrich and Bruno Latour, the term can be defined as follows:

> *Actant:* Whatever acts or shifts actions, action itself being defined by a list of performances through trials; from these performances are deduced a set of competences with which the actant is endowed; the fusion point of a metal is a trial through which the strength of an alloy is defined; the bankruptcy of a company is a trial through which the faithfulness of an ally may be defined; an actor is an actant endowed with a character (usually anthropomorphic) (Akrich and Latour 1992, p. 259, emphasis in original).

In science and technology studies, the notion of actant is frequently applied to technological artifacts, as also Christian Illies and Anthonie Meijers explain:[7]

> [A]ctor-network theory states that technological artefacts 'act' and that together with human agents they are grouped in the same category of 'actants' (this is the principle of generalised symmetry) (Illies and Meijers 2009, p. 420).

Therefore, I propose that memes can also be considered as actants in the sense of actor-network theory. More precisely, there is a relatively new strand of literature dealing with so-called *non-corporeal actants* (NCA) (Hartt 2019). Indeed, Christopher Hartt, who has developed NCA theory, explicitly acknowledges potential complementarities between memetics and NCA theory (see Hartt 2019, especially

---

[6] According to Gibson (2015), the "*affordances* of the environment are what it *offers* the animal, what it *provides* or *furnishes*, either for good or ill. The verb *to afford* is found in the dictionary, but the noun *affordance* is not. I have made it up. I mean by it something that refers to both the environment and the animal in a way that no existing term does. It implies the complementarity of the animal and the environment" (p. 119, italics in original). As Letiche et al. (2011, p. 171) clarify: "Affordances are possibilities presented to consciousness—possibilities for action that the world just seems to be begging us to utilize. The most common example is the chair that is so inviting that it seems to be begging us to sit down ... Likewise, cultural and social circumstances afford applause or booing, joy or sorrow, energy or passivity."

[7] Moreover, one might argue that this sentiment is, to some extent, also reflected in Kevin Kelly's (2010) book *What Technology Wants* (although Kelly does not use the term actant).

his Chap. 13 on *Controversies in NCA theory*). Nevertheless, as already mentioned above, all of these suggestions about potential synergies and their implications should be explored in more detail elsewhere.[8] Let us, for now, bring the discussion about agency of memes to an end with a perfectly fitting quote by Philipp Kneis that also nicely bridges to the next section on creativity and novelty:

> No one, or only a very few people, would actually want to believe they are not in control of things. How could an imaginary unit of cultural replication be controlling my thoughts and actions? Am I not in control? 'I' actually do feel in control. But the 'I' is a construction of the mind, a helpful, but not an unproblematic one. I … am currently typing words that are visible on my computer screen, later on paper. I am thinking these things, typing them … [However,] I did not invent the English language. I did not invent the computer. Neither did I build the computer. I did not invent memetics, … I did not invent academic writing but have seen others do it and mimic their style in part. I do cite my sources, at least the ones I know about. I might be citing things unwillingly. Some things are just 'discourse,' they are 'in the air.' … So, what exactly am I authoring? This text here is a distinct thing. But it is composed of little things not my own. All I can do is create new combinations, thus, pardon the Star Trek reference, creating one instance amongst an infinity of diversity in an infinity of combinations. I am an agent of selection: I choose whom to cite and whom not, whom to talk about and whom not. How free am I in this choice? … My, or any other author's choice, is limited (Kneis 2010, p. 38).

7.2 Everything is a Remix? A Note on Creativity and Novelty[9]

Based on the above quote, it is not very hard to see that the issue of agency is also closely related to the notion of creativity—especially in the sense of a power or capability to create novelty. Consequently, it should come as no surprise that creativity is of interest to myriad scientific disciplines from psychology to innovation economics to cosmology. Depending on the discipline, however, the term creativity can mean very different things (e.g., Abel 2006; Andersson 1997; Boden 1994a, 2004, 2011; Csikszentmihalyi 1996; Frigotto 2018; Gaut and Kieran 2018; Kauffman 2016; Kaufman and Sternberg 2010; Rickards et al. 2009; Sternberg 1999). Although we cannot delve too deeply into the various meanings, it should at least be noted that the concept presumably originated in cosmology. More precisely, it has been argued that the notion of creativity has its roots in the *process philosophy* of Alfred North Whitehead (1978), who is sometimes even said to have coined the term.[10] According to Whitehead's process philosophy, "creativity" describes the fact that we are living in an evolutionary universe of active "innovations," continuously producing novelty.

---

[8] For example, it could be relevant to discuss the differences and similarities between i-memes and e-memes as actants and/or affordances.

[9] I would like to thank Michael Schramm for his contribution to this section by providing me with the paragraph on Whitehead.

[10] For example, see Kristeller (1983, p. 105), Ford (1987, p. 179), or Meyer (2005, p. 2) on this issue.

Hence, the term creativity addresses the momentum of inevitably proceeding into the new—the perpetual process, permanent activity, relentless "ongoingness" (Sherburne 1981, p. 218)—thus pointing to the fact that everything is in a state of flux (cf. Heraclitus' *panta rhei*). In Whitehead's approach, creativity is the fundamental or ultimate principle of the cosmos: "In all philosophic theory there is an ultimate ... In the philosophy of organism, this ultimate is termed 'creativity'" (Whitehead 1978, p. 7). Put differently: "'Creativity' is the universal of universals characterizing ultimate matter of fact ... 'Creativity' is the principle of *novelty*" (Whitehead 1978, p. 21, italics in original).[11]

A memetic understanding of creativity seems perfectly compatible with this Whiteheadian paradigm. On a more general note, however, it should be acknowledged that there already exists ample literature on the merits and limits of Darwinian approaches to creativity that cannot possibly be discussed in detail in this section (e.g., Dietrich and Haider 2015; Gabora and Kauffman 2016; Gabora 2018; Kronfeldner 2011; Simonton 2003; Wagner 2019, to name but a few). Hence, several problems associated with the memetic perspective on creativity have already been addressed elsewhere or are the topics of ongoing debates.

Quite unsurprisingly, given their views on human agency, many memeticists particularly emphasize the *blind* variation part (Campbell 1960, 1974) of the Darwinian "evolutionary algorithm" (e.g., cf. Beinhocker 2006) also in the context of human "ingenuity." In other words, the ideas of "design without a designer" (Dennett 1995) or "competence without comprehension" (Dennett 2017) are transferred from evolutionary biology to novelty creation in the cultural sphere.[12] For example, as Blackmore puts it:

> Many people seem to think that imitation is a crude and blindly mechanistic process that is the antithesis of human creativity, which is conscious and purposive. Theirs is indeed a very different view from my own, and entirely misses the point that evolutionary processes are creative—arguably the only creative processes on the planet. The alternative ... is that just as biological creations come about through natural selection, so human artistic, literary, and scientific creations come about through memetic selection. In both cases the creative force is the evolutionary algorithm. Human achievements are no less creative for that, but our own role has to be seen as that of the clever imitation machine taking part in this new evolutionary process, rather than a conscious entity who can stand outside of it and direct it (Blackmore 2000, pp. 28–29).

Since we already addressed the issue of agency in the previous section, let us focus on the implications for novelty creation, here. According to Blackmore, "human creativity emerges from the human capacity to store, vary, and select memes" (Blackmore 2010, p. 269, with reference to Blackmore 2007). More precisely, new memes are created through "variations on old ones and new combinations created by the compli-

---

[11] Philosophically inclined readers may also want to refer to Schindler's (1973) discussion of creativity and actuality in Whitehead in comparison to Aristotle's (prime) matter, and Aquinas' act of existence (esse).

[12] However, see also Witt (2009), for an enlightening discussion on novelty creation and a critique of the "blind" variation aspect.

cated processes inside a clever thinking brain" (Blackmore 1999, p. 239). Blackmore (2002) specifies this position in her response to Jahoda (2002):

> On this view truly novel human inventions ... come about as a result of a vast evolutionary process ... Variation is introduced either by degradation (such as forgetting or misremembering) or by recombination of memes to produce new ones. Some people seem to interpret this as meaning that there is no true human creativity and no true novelty. This is a bad misunderstanding ... [More precisely,] the creative force is now seen to be the evolutionary process rather than an individual designer ... This is an inevitable and algorithmic process but we humans are not 'passive' ... Indeed our intelligence, capacity for making choices, and active social life are all part of the copying environment in which memes compete (Blackmore 2002, pp. 70–71).

Leaving the somewhat ambiguous role of the "capacity for making choices" in Blackmore's comment aside at this point, for the most part, the argument is closely related to and compatible with the works of Brian Arthur (2007), George Basalla (1988), Joseph Schumpeter (2006), who famously elaborates on "new combinations," and Gabriel Tarde (1890), who writes:

> [L]et us not forget that every invention and every discovery consists in the interference in somebody's mind of certain old pieces of information that have generally been handed down by others (Tarde 1903, p. 382).

Essentially, the memetic perspective can be vividly—and with a fitting ironic side note—summarized in the words of John Ball:

> Where do new memes come from? They arise by combination of parts of some old memes in new ways with perhaps just a jot of originality (mutation). In some fields copying from the work of another is called plagiarism, copying from two or three others is called dull, but copying from many others is called research ... New memes, then, are mostly made of bits and pieces of old memes that have proven to be at least partly successful in the prevailing environment (Ball 1984, p. 148).

There are, however, at least two central issues that need clarification. The first central issue I want to address is that one could get the impression from some of the above quotes that a memetic understanding of creativity reduces the complexity of novelty creation to random mutations, copying errors, and the recombination of memes in an individual's mind.[13] This can happen especially when we cling too much to the supposed analogies between genes and memes. As Aaron Lynch puts it:

> [I]t appears that both the creative formation of new ideas and the development of co-adapted combinations of ideas are affected by the epidemiology of precursors and combinations. This suggests that creativity can be studied not only as an individual phenomenon, but also as a population phenomenon (Lynch 2003, p. 10).

In other words, and as we have also seen in Chaps. 5 and 6 above, one should not neglect the complexity of network effects as addressed by the vast literature on *innovation systems* (e.g., Rakas and Hain 2019) and *innovation networks* (e.g., Buchmann

---

[13] Moreover, when we talk about novelty creation, we should also be aware of the differences between subjective novelty and objective novelty (Witt 2009) or psychological and historical creativity (Boden 2004).

and Pyka 2012; Pyka and Scharnhorst 2009). Another case in point for a more "networked" approach to creativity is Mihály Csíkszentmihályi's so-called *systems model of creativity* (e.g., Csikszentmihalyi 2014a, b) that provides a valuable perspective on creativity in its own right.[14]

The second central issue is that, by focusing too much on *blind* variation, the cases of "radical invention by deliberate human design," as Arthur (2007, p. 275) calls it, appear to be neglected.[15] Moreover, as Bart Nooteboom reminds us, "[r]adical, reconstructive novelty, as opposed to incremental, cumulative novelty, entails uncertainty … it is unpredictable" (Nooteboom 1999, p. 128). On a related note, Margaret Boden famously argues that there are other kinds of creativity beyond so-called *combinational* creativity, namely *exploratory* and *transformational* creativity (e.g., Boden 1994b, 2004, 2011).[16] Consequently, by attempting to color all kinds of innovation and creativity with the same ("panmemetic") brush, important differences and, thus, the explanatory potential of memetics may become concealed. To quote Boudry and Hofhuis again (2018, p. 159), "[i]f selfish memes explain everything, they explain nothing." Nevertheless, there are many cases where memetics helps us to remember that we may not be the geniuses we tend to think we are (as is also suggested by the many instances of serendipity mentioned by Johnson 2010).

Finally, to take up the title of this section, we turn our attention to the notion of *remix*, which has gained some popularity not least due to the well-known video series *Everything is a Remix* by Kirby Ferguson (2015) and his accompanying TED Talk (Ferguson 2012). Ferguson's entertaining but at the same time well-researched videos are also aimed at debunking various myths related to creativity and innovation.[17] In the end, Ferguson (2015) argues, it all comes down to the three fundamental principles of remix: *copy, transform, combine*. Although Ferguson's (2012, 2015) examples have arguably been carefully chosen to support his case, we can see a striking resemblance between the remix principle and the memetic perspective on creativity: copying or imitation, transformation or mutation, and (re-)combination are also central explanatory variables in memetics.

---

[14] At the same time, it may not be too difficult to connect Csikszentmihalyi's systems model to memetics, as the author frequently refers to memes in his own writings, anyway (e.g., Csikszentmihalyi 1993, 1998; Csikszentmihalyi and Massimini 1985).

[15] However, see Wagner (2019, Chap. 7) for clarifications of common misunderstandings regarding "blind variation."

[16] To take up Boden's (2006, p. 25) summary: "Combinational creativity involves unfamiliar combinations of familiar ideas. … Exploratory creativity involves the exploration of some accepted style of thinking, or conceptual space. … Transformational creativity involves the alteration of one or more dimensions of that space, so that structures can now be generated which were previously impossible."

[17] For related arguments in written form, see Berkun (2010) and Lessig (2008).

To conclude this section, I recommend (thoughtful) adoption of Ferguson's (2015) remix principle as a heuristic for memetic approaches to creativity. However, while I tend to agree with the centrality of these three remix principles,[18] I would, at the same time, caution against underestimating the importance of collaborative efforts (e.g., co-design and co-creation), *stigmergy* (Heylighen 2016), and participatory culture for creativity and memetic evolution (e.g., Aharoni 2019; Secretan 2013; Voigts 2017; Wiggins 2017). Finally, because the previous studies (especially Chaps. 5 and 6) have put much emphasis on the *diffusion* aspect, that is, the transmission and assimilation stages of the replication loop (Heylighen and Chielens 2009), I propose to systematically explore conceptual complementarities and potential synergies between remix and established approaches to knowledge *creation* in evolutionary economics and organization science as an avenue for future research. For example, future studies could inquire into the role and nature of remix with an eye to "absorptive capacities" (e.g., Cohen and Levinthal 1990), "dynamic capabilities" (e.g., Teece and Pisano 2004; Teece et al. 1997), or the differences between "exploration and exploitation" approaches (e.g., March 1991). Ultimately, creativity remains a complex and multidimensional phenomenon, so that many of the open issues must be left to a variety of experts from other disciplines to address (e.g., neuroscientists, philosophers of mind, quantum physicists, etc.).

7.3 Memethics: Normative Implications

Before we turn to the conclusion of this book in Chap. 8, we still have to deal with one of the most important issues of all: *ethics*. More precisely, this section addresses the normative implications of econometrics. As with the previous two sections, however, we should not expect this passage to yield fully-fledged solutions or a comprehensive moral-philosophical essay. Most of the issues touched upon here can be seen as special cases of the more general discussions about *evolutionary ethics* (e.g., Boniolo and de Anna 2006; Bayertz 1993; Campbell 1979; Cela-Conde and Ayala 2004; Hodgson 2014; Illies 2013; James 2011; Levy 2010; Mizzoni 2017; Mohr 2014; Sober and Wilson 1998; Stewart-Williams 2015; Wuketits 1993, 2009).

One may wonder why I write that we "have to" deal with ethics (in the first sentence of this section). Arguably, memetics is bursting with implicit normative consequences that simply cannot be ignored with a clear conscience. For example, if we carry memetics to extremes, we (humans) are simply the slaves of "our" selfish replicators, so that we are neither responsible for our actions nor able to change anything for the better. Well then, should we just sit back, relax, and watch what evolution has in petto for us? While this all relates back to the complex issue of

---

[18]It should go without saying that my endorsement of the remix principle does not mean that I deny the existence of (subjective) novelty or the introduction of "novelty" from one (sub-)system to another by means of co-evolution or so-called *promotion* (Almudi and Fatas-Villafranca 2018). I simply argue that, from an overall system (or objective) point of view, novelty must always arise from what already exists, otherwise we would have some kind of *creatio ex nihilo*.

agency, it may help to see that Dawkins himself does not seem to be convinced that we are just helpless puppets of our genes and memes:

> We are built as gene machines and cultured as meme machines, but we have the power to turn against our creators. We, alone on earth, can rebel against the tyranny of the selfish replicators (Dawkins 2016, p. 260).

In other words, it is reasonable to assume that we can influence the evolutionary trajectories of both genes and memes to a certain (situation-dependent) degree, which is in line with the continuum perspective on agency mentioned in Sect. 7.1 above. What does this mean for morality and the economy, then? Here, it is important to remember from the introductory Chap. 1 that cultures can be regarded as "economic moral cultures" (*ökonomische Moralkulturen*), according to Michael Schramm (2008). In other words, we know from the literature that cultural influences on what people consider to be right or wrong also matter for business and economic activity, as evidenced (not just) by the textbook example of (the perception of) effective leadership practices in different cultural contexts (e.g., Chhokar et al. 2008; House et al. 2014; Schein 2017; Schlaile 2012).

If we break these insights down to the replicator level, we might adopt the term *moral meme* proposed by Steve Stewart-Williams (2015) to denote those culturally evolved moral codes that are the central elements (or core values) of an individual's moral system. As Robert Boyd and Peter Richerson explain in an article from the mid-nineties, when they still used the term meme:

> We refer to alternative culturally transmitted items of information as 'memes.' Norms, on this view, are those memes which influence standards of behavior. Since memes are communicated from one person to another, individuals sample from and contribute to an evolving meme pool ... To understand why people behave as they do in a particular environment, we must know the nature of the skills, beliefs, attitudes, and values that they have acquired from others by cultural inheritance. To do this we must account for the processes that affect cultural variation as individuals acquire memes, use the acquired information to guide behavior, and act as models for others (Boyd and Richerson 1994, p. 74).

In other words, what Dawkins (2006, Chap. 7) calls "the changing moral *Zeitgeist*" (italics in original) is essentially an instance of the cultural evolution of "moral memes." In their paper, Boyd and Richerson (1994) focus particularly on two cultural evolutionary forces: *biased transmission* and *natural selection*. Although we know from the literature on cultural evolution that there are also other factors involved (e.g., Boyd and Richerson 2005), it is probably no exaggeration to state that these two factors are crucial for the evolution of economic moral cultures and also important from a memetic point of view (in fact, Chap. 5 may be read as an attempt to combine elements of biased transmission and natural selection). Interestingly, the natural selection of "morals" was already discussed in 1892 in a paper by Samuel Alexander:[19]

---

[19] Note how this also supports what we have learned about the remix principle in the previous section.

> Now, in calling natural selection in ethics a struggle of ideals, I am trying to represent shortly and compendiously this possibility of life, growth, and alteration of people's minds through the spontaneous or artificial production of new ideas (Alexander 1892, p. 420).

And to highlight the similarity with memetics even more:

> The principle that I have been trying to explain, then, is, that ideals of conduct are established, that distinctions of good and bad are created and changed by a process of struggle,—an internecine struggle which is identical with that of natural selection, but works upon minds, not upon bodies, or rather it need not, though it sometimes must, work upon bodies as well (Alexander 1892, p. 426).

Whenever we apply Darwinian ideas to social phenomena, though, we should not forget to mention the issues associated with so-called *Social Darwinism* (e.g., Beck 2013; Hodgson 2004b; Wuketits 1993). In the nineteenth century, proponents of Social Darwinism often unjustifiably conflated the outcomes of evolutionary processes with value judgments that were arguably used to promote war, racism, eugenics, fascism, and laissez-faire capitalism—in bold and simple terms, survival of the fittest was mistaken for survival of the most valuable or most precious. While this is a prime example of the *naturalistic fallacy*, memetics can actually be commended for avoiding this mistake because most proponents of memetics have always insisted that those memes which survived the competition for attention against other memes are not necessarily true or good or right; they just had "the right" selection environment.[20] Put differently, most memeticists seem to be aware of the fact that *one does not simply* derive normative from descriptive propositions.

What do we take from this discussion? In plain terms, our morality is influenced by cultural evolution.[21] In less plain terms, moral memes are part of the schemata that constitute a *worldview* or *paradigm*. To draw from a definition we already used elsewhere:

> A paradigm ... can be defined as a complex set of assumptions, concepts, values, and practices that constitute a worldview for the community that shares them ... Hence, paradigms span a *bounded performative space* within which certain actions or practices are regarded as possible, reasonable, legitimate, and important, while others are excluded as being impossible, illegitimate, unreasonable, and unimportant ... This performative space actuates but also bounds the emergence and development of practices (Schlaile et al. 2017, p. 4, italics in original).

In this regard, I propose that economemetics can help us to generate more insights into the evolution of what various sustainability scientists call "normative knowledge" (e.g., see Urmetzer et al. 2018, for details). Put differently, if I accept the premise that we can, to a certain degree, influence the cultural evolutionary process, and if my intention as a researcher is to explain or even actively contribute to what Maja

---

[20]On a related note, one may argue that various innovation scholars and politicians have committed the naturalistic fallacy by implicitly regarding innovation *per se* as desirable (e.g., see Schlaile et al. 2018, and references therein).

[21]For the sake of brevity, we will not go into debates about the implications of this statement for the position of *moral realism* (e.g., Sayre-McCord 2015), here.

Göpel calls *The Great Mindshift* (Göpel 2016), I should perhaps take the scientific literature on memetics more seriously.

This view is prominently supported by Sandra Waddock (2015, 2016, 2019), who at the same time underlines the relationship between memetics and framing:

> Memes ... form the basis of what Lakoff (2014) calls frames or the narratives and stories that shape what and how we believe the world operates, what is important, and where attention needs to be placed (and 'how things work here' in companies). Memes thus provide the foundation of how people perceive the world, their arguments, and the visions about constructing the future, among other things. In a change process, memes shape how to make sense of what is changing (Waddock 2019, p. 935, with reference to Lakoff 2014).

What this discussion essentially tells us is that there is no such thing as "mem*ethics*" in the sense of a memetic theory of what is right or wrong. However, what memetics may eventually be able to contribute to—especially in connection with findings from related disciplines—is an explanation of the differential *cultural evolution of right and wrong* in terms of biased transmission and natural selection of moral memes.

In summary, it can be said that econememetics has several normative implications that may be taken up in future discussions. First, I have argued that a nuanced perspective on the complexity of agency leads to a rejection of memetic determinism and, thus, allows for a degree of responsibility. However, the morality of economic actors is bounded by the differential cultural evolution (one may be tempted to say "speciation") of moral memes and other non-memetic factors. Second, econememetics also leads to a rejection of the view that successful replication of worldviews or paradigms makes them morally superior to other worldviews. Consequently, memetics alone cannot provide normative propositions, but it can help to explain and potentially influence the evolution of normative knowledge in a direction desired by the researcher. Finally, I suggest that econememetics may help to close some research gaps in the literature on the normative dimension of sustainability transitions by serving as a heuristic and as a link to related concepts, but it cannot be the *overarching* theoretical framework in this regard.[22]

References

Abel, G. (Ed.). (2006). *Kreativität: XX. Deutscher Kongreß für Philosophie: Kolloquienbeiträge*. Hamburg: Meiner.

Aharoni, T. (2019). When high and pop culture (re)mix: An inquiry into the memetic transformations of artwork. *New Media & Society*. https://doi.org/10.1177/1461444819845917.

Akrich, M., & Latour, B. (1992). A summary of a convenient vocabulary for the semiotics of human and nonhuman assemblies. In W. E. Bijker & J. Law (Eds.), *Shaping technology/ building society: Studies in sociotechnical change*. Cambridge: MIT Press.

Alexander, S. (1892). Natural selection in morals. *International Journal of Ethics, 2*(4), 409–439.

---

[22] For a related and much more extensive argument in this direction, see, for example, van den Bergh (2018) or Waring (2010).

7 General Discussion: Economemetics and Agency, Creativity, and Normativity

Almudi, I., & Fatas-Villafranca, F. (2018). Promotion and coevolutionary dynamics in contemporary capitalism. *Journal of Economic Issues, 52*(1), 80–102. https://doi.org/10.1080/00213624.2018.1430943.

Andersson, Å. E., & Sahlin, N.-E. (Eds.). (1997). *The complexity of creativity*. Dordrecht: Springer.

Archer, M. S. (1996). *Culture and agency: The place of culture in social theory* (2nd ed.). Cambridge: Cambridge University Press.

Arthur, W. B. (2007). The structure of invention. *Research Policy, 36*(2), 274–287. https://doi.org/10.1016/j.respol.2006.11.005.

Aunger, R. (2002). *The electric meme: A new theory about how we think*. New York: Free Press.

Ball, J. A. (1984). Memes as replicators. *Ethology and Sociobiology, 5*, 145–161. https://doi.org/10.1016/0162-3095(84)90020-7.

Barker, C. (2012). *Cultural studies: Theory and practice* (4th ed.). Los Angeles: Sage.

Basalla, G. (1988). *The evolution of technology*. Cambridge: Cambridge University Press.

Bayertz, K. (Ed.). (1993). *Evolution und Ethik*. Stuttgart: Reclam.

Beck, N. (2013). Social Darwinism. In M. Ruse (Ed.), *The Cambridge encyclopedia of Darwin and evolutionary thought* (pp. 195–201). Cambridge: Cambridge University Press.

Beinhocker, E. D. (2006). *The origin of wealth: Evolution, complexity, and the radical remaking of economics*. Boston: Harvard Business School Press.

Berkun, S. (2010). *The myths of innovation*. Sebastopol: O'Reilly.

Bijker, W. E., & Law, J. (Eds.). (1992). *Shaping technology/building society: Studies in sociotechnical change*. Cambridge: MIT Press.

Blackmore, S. (1999). *The meme machine*. Oxford: Oxford University Press.

Blackmore, S. (2000). The meme's eye view. In R. Aunger (Ed.), *Darwinizing culture: The status of memetics as a science* (pp. 25–42). Oxford: Oxford University Press.

Blackmore, S. (2002). A response to Gustav Jahoda. *History of the Human Sciences, 15*(2), 69–71. https://doi.org/10.1177/0952695102015002131.

Blackmore, S. (2005). Can memes meet the challenge? Susan Blackmore on Greenberg and on Chater. In S. Hurley & N. Chater (Eds.), *Perspectives on imitation: From neuroscience to social science* (Vol. 2, pp. 409–411). Imitation, human development, and culture. Cambridge: The MIT Press.

Blackmore, S. (2007). Memes, minds, and imagination. In I. Roth (Ed.), *Imaginative minds* (pp. 61–78). Proceedings of the British Academy. Oxford: Oxford University Press.

Blackmore, S. (2010). Memetics does provide a useful way of understanding cultural evolution. In F. J. Ayala & R. Arp (Eds.), *Contemporary debates in philosophy of biology* (pp. 225–272). Chichester: Wiley-Blackwell.

Blok, A., Farías, I., & Roberts, C. (Eds.). (2020). *The Routledge companion to actor-network theory*. Abingdon: Routledge.

Blute, M. (2010). *Darwinian sociocultural evolution: Solutions to dilemmas in cultural and social theory*. Cambridge: Cambridge University Press.

Boden, M. A. (Ed.). (1994a). *Dimensions of creativity*. Cambridge: MIT Press.

Boden, M. A. (1994b). What is creativity. In M. A. Boden (Ed.), *Dimensions of creativity* (pp. 75–117). Cambridge: MIT Press.

Boden, M. A. (2004). *The creative mind: Myths and mechanisms* (2nd ed.). London: Routledge.

Boden, M. A. (2006). The concept of creativity (Abstract). In G. Abel (Ed.), *Kreativität: XX. Deutscher Kongreß für Philosophie* (p. 25). Hamburg: Meiner.

Boden, M. A. (2011). *Creativity and art: Three roads to surprise*. Oxford: Oxford University Press.

Boniolo, G., & de Anna, G. (Eds.). (2006). *Evolutionary ethics and contemporary biology*. Cambridge: Cambridge University Press.

Boudry, M., & Hofhuis, S. (2018). Parasites of the mind. Why cultural theorists need the meme's eye view. *Cognitive Systems Research, 52*, 155–167. https://doi.org/10.1016/j.cogsys.2018.06.010.

Boyd, R., & Richerson, P. J. (1994). The evolution of norms: an anthropological view. *Journal of Institutional and Theoretical Economics (JITE), 150*(1), 72–87.

Boyd, R., & Richerson, P. J. (2005). *The origin and evolution of cultures*. Oxford: Oxford University Press.

Buchmann, T., & Pyka, A. (2012). Innovation networks. In M. Dietrich & J. Krafft (Eds.), *Handbook on the economics and theory of the firm* (pp. 466–482). Cheltenham: Edward Elgar.

Burns, T. R., & Dietz, T. (1992). Cultural evolution: Social rule systems, selection and human agency. *International Sociology*, 7(3), 259–283. https://doi.org/10.1177/026858092007003001.

Callinicos, A. (2004). *Making history: Agency, structure, and change in social theory* (2nd ed.). Leiden: Brill Academic Publishers.

Campbell, D. T. (1960). Blind variation and selective retention in creative thought as in other knowledge processes. *Psychological Review*, 67(6), 380–400. https://doi.org/10.1037/h0040373.

Campbell, D. T. (1974). Evolutionary epistemology. In P. A. Schilpp (Ed.), *The philosophy of Karl Popper* (pp. 413–463). Lasalle: Open Court.

Campbell, D. T. (1979). Comments on the sociobiology of ethics and moralizing. *Behavioral Science*, 24(1), 37–45. https://doi.org/10.1002/bs.3830240106.

Cela-Conde, C. J., & Ayala, F. J. (2004). Evolution of morality. In F. M. Wuketits & C. Antweiler (Eds.), *Handbook of evolution* (Vol. 1, pp. 171–189). The evolution of human societies and cultures. Weinheim: Wiley.

Chhokar, J. S., Brodbeck, F. C., & House, R. J. (Eds.). (2008). *Culture and leadership across the world: The GLOBE book of in-depth studies of 25 societies*. New York: Taylor & Francis.

Cohen, W. M., & Levinthal, D. A. (1990). Absorptive capacity: A new perspective on learning and innovation. *Administrative Science Quarterly*, 35(1), 128–152.

Csikszentmihalyi, M. (1993). *The evolving self: A psychology for the third millennium*. New York: HarperCollins.

Csikszentmihalyi, M. (1996). *Creativity: Flow and the psychology of discovery and invention*. New York: Harper.

Csikszentmihalyi, M. (1998). Self and evolution. *The NAMTA Journal*, 23(1), 205–233.

Csikszentmihalyi, M. (2014a). *The systems model of creativity*. Dordrecht: Springer.

Csikszentmihalyi, M. (2014b). The systems model of creativity and its applications. In D. K. Simonton (Ed.), *The Wiley handbook of genius* (pp. 533–545). Chichester: Wiley.

Csikszentmihalyi, M., & Massimini, F. (1985). On the psychological selection of bio-cultural information. *New Ideas in Psychology*, 3(2), 115–138. https://doi.org/10.1016/0732-118X(85)90002-9.

Dawkins, R. (2006). *The God delusion*. London: Bantam Press.

Dawkins, R. (2016). *The selfish gene* (40th anniversary ed.). Oxford: Oxford University Press.

Dennett, D. C. (1987). *The intentional stance*. Cambridge: MIT Press.

Dennett, D. C. (1995). *Darwin's dangerous idea: Evolution and the meanings of life*. New York: Simon & Schuster.

Dennett, D. C. (2017). *From bacteria to Bach and back: The evolution of minds*. New York: W.W. Norton & Company.

Dietrich, A., & Haider, H. (2015). Human creativity, evolutionary algorithms, and predictive representations: The mechanics of thought trials. *Psychonomic Bulletin & Review*, 22(4), 897–915. https://doi.org/10.3758/s13423-014-0743-x.

Dietz, T., & Burns, T. R. (1992). Human agency and the evolutionary dynamics of culture. *Acta Sociologica*, 35(3), 187–200. https://doi.org/10.1177/000169939203500302.

Ferguson, K. (2012). Embrace the remix. *TEDGlobal*. Retrieved from https://www.ted.com/talks/kirby_ferguson_embrace_the_remix.

Ferguson, K. (2015). Everything is a remix (remastered). *Vimeo*. Retrieved from https://vimeo.com/139094998.

Ford, L. S. (1987). Creativity in a future key. In R. C. Neville (Ed.), *New essays in metaphysics* (pp. 179–197). Albany: State University of New York Press.

Frigotto, M. L. (2018). *Understanding novelty in organizations: A research path across agency and consequences*. Cham: Springer.

Fuchs, S. (2001). Beyond agency. *Sociological Theory, 19*(1), 24–40. https://doi.org/10.1111/0735-2751.00126.
Gabora, L. (2018). The creative process of cultural evolution. In A. K.-Y. Leung, L. Y.-Y. Kwan, & S. Liou (Eds.), *Handbook of culture and creativity: Basic processes and applied innovations* (pp. 33–57). New York: Oxford University Press.
Gabora, L., & Kauffman, S. (2016). Toward an evolutionary-predictive foundation for creativity. *Psychonomic Bulletin & Review, 23*(2), 632–639. https://doi.org/10.3758/s13423-015-0925-1.
Gaut, B. N., & Kieran, M. (Eds.). (2018). *Creativity and philosophy*. London: Routledge.
Gibson, J. J. (2015). *The ecological approach to visual perception* (Classic ed.). New York: Taylor & Francis; (2nd ed., 1986).
Göpel, M. (2016). *The great mindshift: How a new economic paradigm and sustainability transformations go hand in hand*. Berlin: Springer.
Hartt, C. M. (Ed.). (2019). *Connecting values to action: Non-corporeal actants and choice*. Bingley: Emerald Publishing.
Herrmann-Pillath, C. (2013). *Foundations of economic evolution: A treatise on the natural philosophy of economics*. Cheltenham: Edward Elgar.
Herrmann-Pillath, C. (2018). *Grundlegung einer kritischen Theorie der Wirtschaft*. Marburg: Metropolis.
Heylighen, F. (2016). Stigmergy as a universal coordination mechanism I: Definition and components. *Cognitive Systems Research, 38*, 4–13. https://doi.org/10.1016/j.cogsys.2015.12.002.
Heylighen, F., & Chielens, K. (2009). Evolution of culture, memetics. In R. A. Meyers (Ed.), *Encyclopedia of complexity and systems science* (pp. 3205–3220). New York: Springer.
Hodgson, G. M. (2004a). Social Darwinism in anglophone academic journals: A contribution to the history of the term. *Journal of Historical Sociology, 17*(4), 428–463. https://doi.org/10.1111/j.1467-6443.2004.00239.x.
Hodgson, G. M. (2004b). *The evolution of institutional economics: Agency, structure, and Darwinism in American institutionalism*. London: Routledge.
Hodgson, G. M. (2014). The evolution of morality and the end of economic man. *Journal of Evolutionary Economics, 24*(1), 83–106. https://doi.org/10.1007/s00191-013-0306-8.
Hofhuis, S., & Boudry, M. (2019). 'Viral' hunts? A cultural Darwinian analysis of witch persecutions. *Cultural Science Journal, 11*(1), 13–29. https://doi.org/10.5334/csci.116.
House, R. J., Dorfman, P. W., Javidan, M., Hanges, P. J., & Sully De Luque, M. F. (2014). *Strategic leadership across cultures: The GLOBE study of CEO leadership behavior and effectiveness in 24 countries*. Los Angeles: Sage.
Hull, D. L. (1999). Strategies in meme theory - a commentary on Rose's paper: Controversies in meme theory. *Journal of Memetics - Evolutionary Models of Information Transmission, 3*. Retrieved from http://cfpm.org/jom-emit/1999/vol3/hull_dl.html
Illies, C. (2013). Evolutionär erweiterte Ethik. Fünf Thesen zur Bedeutung der Evolutionswissenschaften für die Ethik. In H.P. Weber & R. Langthaler (Eds.), *Evolutionstheorie und Schöpfungsglaube* (pp. 361–382). Göttingen: V&R Unipress.
Illies, C., & Meijers, A. (2009). Artefacts without agency. *The Monist, 92*(3), 420–440. https://doi.org/10.5840/monist200992324.
Ingold, T. (1986). *Evolution and social life*. Cambridge: Cambridge University Press.
Jahoda, G. (2002). The ghosts in the meme machine. *History of the Human Sciences, 15*(2), 55–68. https://doi.org/10.1177/0952695102015002126.
James, S. M. (2011). *An introduction to evolutionary ethics*. Malden: Wiley.
Johnson, S. (2010). *Where good ideas come from: The natural history of innovation*. New York: Riverhead Books.
Kauffman, S. A. (2000). *Investigations*. New York: Oxford University Press.
Kauffman, S. A. (2016). *Humanity in a creative universe*. New York: Oxford University Press.
Kaufman, J. C., & Sternberg, R. J. (Eds.). (2010). *The Cambridge handbook of creativity*. Cambridge: Cambridge University Press.
Kelly, K. (2010). *What technology wants*. New York: Penguin.

Kneis, P. (2010). *The emancipation of the soul: Memes of destiny in American mythological television*. Frankfurt a. M.: Peter Lang.
Kristeller, P. O. (1983). "Creativity" and "tradition". *Journal of the History of Ideas, 44*(1), 105–113. https://doi.org/10.2307/2709307.
Kronfeldner, M. (2011). *Darwinian creativity and memetics*. Durham: Acumen.
Krul, M. (2018). The new institutionalist economic history of Douglass C. North: A critical interpretation. Cham: Palgrave Macmillan/Springer Nature.
Lakoff, G. (2014). *The all new don't think of an elephant! Know your values and frame the debate*. White River Junction: Chelsea Green Publishing.
Latour, B. (2005). *Reassembling the social: An introduction to actor-network-theory*. Oxford: Oxford University Press.
Lessig, L. (2008). *Remix: Making art and commerce thrive in the hybrid economy*. New York: Penguin.
Letiche, H., Lissack, M., & Schultz, R. (2011). *Coherence in the midst of complexity: Advances in social complexity theory*. New York: Palgrave Macmillan.
Levy, N. (Ed.). (2010). *Evolutionary ethics*. Farnham: Ashgate.
Lynch, A. (2003). An introduction to the evolutionary epidemiology of ideas. *The Biological Physicist, 3*(2), 7–14.
March, J. G. (1991). Exploration and exploitation in organizational learning. *Organization Science, 2*(1), 71–87. https://doi.org/10.1287/orsc.2.1.71.
McGrenere, J., & Ho, W. (2000). Affordances: clarifying and evolving a concept. In S. S. Fels & P. Poulin (Eds.), *Proceedings of the Graphics Interface 2000 Conference* (pp. 179–186). San Francisco: Morgan Kaufmann Publishers.
Meyer, S. (2005). Introduction. *Configurations, 13*(1), 1–33. https://doi.org/10.1353/con.2007.0010.
Mizzoni, J. (2017). *Evolution and the foundations of ethics: Evolutionary perspectives on contemporary normative and metaethical theories*. Lanham: Lexington Books.
Mohr, H. (2014). *Evolutionäre Ethik. Schriften der Mathematisch-naturwissenschaftlichen Klasse*, Bd. 25. Wiesbaden: Springer.
Nooteboom, B. (1999). Innovation, learning and industrial organisation. *Cambridge Journal of Economics, 23*(2), 127–150. https://doi.org/10.1093/cje/23.2.127.
North, D. C. (2005). *Understanding the process of economic change*. Princeton: Princeton University Press.
Nye, B. D. (2011). Modeling memes: A memetic view of affordance learning. *Publicly accessible Penn Dissertations, 336*. Retrieved from http://repository.upenn.edu/edissertations/336/
Price, I. (1999). Steps toward the memetic self - a commentary on Rose's paper: Controversies in meme theory. *Journal of Memetics - Evolutionary Models of Information Transmission, 3*. Retrieved from http://cfpm.org/jom-emit/1999/vol3/price_if.html
Pyka, A., & Scharnhorst, A. (Eds.). (2009). *Innovation networks. New approaches in modelling and analyzing*. Berlin: Springer.
Rakas, M., & Hain, D. S. (2019). The state of innovation system research: What happens beneath the surface? *Research Policy, 48*(9), 103787. https://doi.org/10.1016/j.respol.2019.04.011.
Rickards, T., Runco, M. A., & Moger, S. (Eds.). (2009). *The Routledge companion to creativity*. London: Routledge.
Rose, N. (1998). Controversies in meme theory. *Journal of Memetics - Evolutionary Models of Information Transmission, 2*. Retrieved from http://cfpm.org/jom-emit/1998/vol2/rose_n.html
Rosenberg, A. (2017). Why social science is biological science. *Journal for General Philosophy of Science, 48*(3), 341–369. https://doi.org/10.1007/s10838-017-9365-0.
Sayre-McCord, G. (2015). Moral realism. In E. N. Zalta (Ed.), *The Stanford encyclopedia of philosophy*, Spring 2015 ed. Metaphysics Research Lab, Stanford University. Retrieved from https://plato.stanford.edu/archives/fall2017/entries/moral-realism/
Schein, E. H. (2017). *Organizational culture and leadership* (5th ed.). Hoboken: Wiley.

Schindler, D. L. (1973). Creativity as ultimate: Reflections on actuality in Whitehead, Aristotle, and Aquinas. *International Philosophical Quarterly, 13*(2), 161–171. https://doi.org/10.5840/ipq197313224.

Schlaile, M. P. (2012). Global Leadership im Kontext ökonomischer Moralkulturen - eine induktiv-komparative Analyse. *Hohenheimer Working Papers zur Wirtschafts- und Unternehmensethik* Nr. 13. Retrieved from https://theology-ethics.uni-hohenheim.de/fileadmin/einrichtungen/theology-ethics/hhwpwue_13_Schlaile.pdf

Schlaile, M. P., & Ehrenberger, M. (2016). Complexity, cultural evolution, and the discovery and creation of (social) entrepreneurial opportunities: Exploring a memetic approach. In E. S. C. Berger & A. Kuckertz (Eds.), *Complexity in entrepreneurship, innovation and technology research: Applications of emergent and neglected methods* (pp. 63–92). Cham: Springer.

Schlaile, M. P., Mueller, M., Schramm, M., & Pyka, A. (2018). Evolutionary economics, responsible innovation and demand: Making a case for the role of consumers. *Philosophy of Management, 17*(1), 7–39. https://doi.org/10.1007/s40926-017-0054-1.

Schlaile, M. P., Urmetzer, S., Blok, V., Andersen, A. D., Timmermans, J., Mueller, M., et al. (2017). Innovation systems for transformations towards sustainability? Taking the normative dimension seriously. *Sustainability, 9*(12), 2253. https://doi.org/10.3390/su9122253.

Schramm, M. (2008). *Ökonomische Moralkulturen. Die Ethik differenter Interessen und der plurale Kapitalismus*. Marburg: Metropolis.

Schumpeter, J. A. (2006). *Theorie der wirtschaftlichen Entwicklung. Nachdruck der 1. Auflage von 1912. Herausgegeben und ergänzt um eine Einführung von Jochen Röpke und Olaf Stiller*. Berlin: Duncker & Humblot.

Searle, J. R. (1997). *The mystery of consciousness*. New York: New York Review of Books.

Secretan, J. (2013). Stigmergic dimensions of online creative interaction. *Cognitive Systems Research, 21*, 65–74. https://doi.org/10.1016/j.cogsys.2012.06.006.

Sherburne, D. W. (Ed.). (1981). *A key to Whitehead's process and reality*. Chicago: University of Chicago Press; (1st ed., 1966)

Simonton, D. K. (2003). *Origins of genius: Darwinian perspectives on creativity*. Oxford: Oxford University Press.

Sober, E., & Wilson, D. S. (1998). *Unto others: The evolution and psychology of unselfish behavior*. Cambridge: Harvard University Press.

Sternberg, R. J. (Ed.). (1999). *Handbook of creativity*. Cambridge: Cambridge University Press.

Stewart-Williams, S. (2015). Morality: Evolution of. In J. D. Wright (Ed.), *International encyclopedia of the social & behavioral sciences* (pp. 811–818). Amsterdam: Elsevier.

Tarde, G. (1890). *Les lois de l'imitation: Étude sociologique*. Paris: Félix Alcan.

Tarde, G. (1903). *The laws of imitation*. Transl. by E. C. Parsons. New York: Henry Holt.

Teece, D. J., Pisano, G., & Shuen, A. (1997). Dynamic capabilities and strategic management. *Strategic Management Journal, 18*(7), 509–533.

Teece, D., & Pisano, G. (2004). The dynamic capabilities of firms. In C. W. Holsapple (Ed.), *Handbook on knowledge management 2: Knowledge directions* (pp. 195–213). Berlin: Springer.

Urmetzer, S., Schlaile, M. P., Bogner, K., Mueller, M., & Pyka, A. (2018). Exploring the dedicated knowledge base of a transformation towards a sustainable bioeconomy. *Sustainability, 10*(6), 1694. https://doi.org/10.3390/su10061694.

van den Bergh, J. C. J. M. (2018). *Human evolution beyond biology and culture: Evolutionary social, environmental and policy sciences*. Cambridge: Cambridge University Press.

Voigts, E. (2017). Memes and recombinant appropriation: Remix, mashup, parody. In T. M. Leitch (Ed.), *The Oxford handbook of adaptation studies*. https://doi.org/10.1093/oxfordhb/9780199331000.013.16

Waddock, S. (2015). Reflections: Intellectual shamans, sensemaking, and memes in large system change. *Journal of Change Management, 15*(4), 259–273. https://doi.org/10.1080/14697017.2015.1031954.

Waddock, S. (2016). Foundational memes for a new narrative about the role of business in society. *Humanistic Management Journal, 1*(1), 91–105. https://doi.org/10.1007/s41463-016-0012-4.

Waddock, S. (2019). Shaping the shift: Shamanic leadership, memes, and transformation. *Journal of Business Ethics, 155*(4), 931–939. https://doi.org/10.1007/s10551-018-3900-8.

Wagner, A. (2019). *Life finds a way: What evolution teaches us about creativity*. New York: Basic Books.

Waring, T. M. (2010). New evolutionary foundations: Theoretical requirements for a science of sustainability. *Ecological Economics, 69*(4), 718–730. https://doi.org/10.1016/j.ecolecon.2008.10.017.

Whitehead, A. N. (1978). *Process and reality: An essay in cosmology: Gifford lectures delivered in the University of Edinburgh during the session 1927–28* (corrected ed.) (D. R. Griffin & D. W. Sherburne, Eds.). New York: Free Press; (1st ed., 1929).

Wiggins, B. E. (2017). Navigating digital culture: Remix culture, viral media, and internet memes. In L. Gómez Chova, A. López Martínez, & I. Candel Torres (Eds.), *INTED2017 Proceedings* (pp. 368–374). IATED.

Witt, U. (2009). Propositions about novelty. *Journal of Economic Behavior & Organization, 70*(1–2), 311–320. https://doi.org/10.1016/j.jebo.2009.01.008.

Wuketits, F. M. (1993). Moral systems as evolutionary systems: Taking evolutionary ethics seriously. *Journal of Social and Evolutionary Systems, 16*(3), 251–271. https://doi.org/10.1016/1061-7361(93)90035-P.

Wuketits, F. M. (2009). Charles Darwin and modern moral philosophy. *Ludus Vitalis, 17*(32), 395–404.

Chapter 8
Conclusion and the Way(s) Forward

Michael P. Schlaile

Abstract This chapter concludes the book by explicitly stating some general limitations and by providing a summary and an outlook. The four general limitations acknowledged here are conceptual ambiguity, under-representation of creativity (or remix), neglect of co-evolution, and the unresolved question of how far the analogies between cultural and biological evolution should go. The chapter also summarizes the book's core findings and highlights the "five i" of economemetics: imitation, information, instruction, innovation, and interconnection. Finally, the chapter invites future research on philosophical implications of economemetics, overlaps and connections with framing theory, semiotics, organizational institutionalism, and alternative economic disciplines, as well as more in-depth inquiries into our understanding of information.

8.1 General Limitations

While the specific limitations of the particular studies and their methodology have already been addressed in the respective chapters and will not be repeated here, there are four rather general limitations to economemetics as presented in this book that should be acknowledged.

The first and probably most obvious—but nevertheless serious—limitation is that we are dealing with a myriad of ambiguous notions and terms, or, what Walter Bryce Gallie (1956) calls *essentially contested concepts*. Many of the central issues studied in this book arguably fall into this category. Examples include culture, diffusion, economy, evolution, imitation, information, institution, knowledge, etc. For instance, due to the simple fact that culture can have so many different meanings (e.g., Baldwin et al. 2006; Kroeber and Kluckhohn 1952), the elements or units of culture—memes—will remain ambiguous as well (remember the sample of meme definitions in Chap. 3, Appendix A).

M. P. Schlaile (✉)
Department of Innovation Economics, University of Hohenheim, Stuttgart, Germany
e-mail: schlaile@uni-hohenheim.de

© The Author(s), under exclusive license to Springer Nature Switzerland AG 2021
M. P. Schlaile (ed.), *Memetics and Evolutionary Economics*, Economic Complexity and Evolution, https://doi.org/10.1007/978-3-030-59955-3_8

The second general limitation of this book is that the aspect of creativity (or remix) has not received much attention in the studies, which have focused either on the diversity and interdependence of a complex network at a particular point in time (Chap. 4) or on the diffusion (or biased transmission) and selection of memes/knowledge units (Chaps. 5 and 6) via static networks. In other words, the incorporation of other relevant aspects of Darwinian evolutionary dynamics in terms of novelty creation (or "genealogy" of memes instead of "epidemiology") is still improvable.

The third general limitation is that the *co-evolution* of different (sub-)systems such as markets, organizations, nature, and political systems has been neglected to make the results of the studies interpretable and to have a clear focus. Nevertheless, as the literature on co-evolution suggests (e.g., Abatecola et al. 2020; Almudi and Fatas-Villafranca 2018; Breslin 2016; Durham 1991), systemic reciprocity and the interdependencies between replicators and interactors within and between different sub-systems are important characteristics of the evolutionary economic process that should not be oversimplified. Put differently, and to link back to Christian Illies' (2010) five *Types* used in Chap. 1 to position this book, it is clear that by focusing on Type D (while using an overall Type E evolutionary perspective), one tends to disregard the role of general boundary conditions (Type A), the sociobiological foundations (Type B), and the positive impacts of culture on human well-being as well as co-evolution (Type C). At this point, it should not be forgotten, however, that the latter two general limitations are perfectly justifiable in terms of the trade-off between sufficient complexity and interpretability of a study. After all, no model is perfect.

A fourth general limitation of this book is an issue that all Darwinian social sciences face, namely the question of how far the analogy to biological evolution can and should be pushed. I have frequently argued for a rather relaxed interpretation of the similarities between genes, memes, and viruses; yet, as Alex Rosenberg writes:

> The issue is serious for Darwinian social science. That natural selection is an attractive metaphor in the description of human affairs is both unexceptional and uninteresting. The issue is whether it is more than a metaphor. Are all social processes literally, actually, really matters of blind variation and environmental filtration? That is the issue (Rosenberg 2017, p. 355).

While the second and third limitation can potentially be resolved by extending the scope of the studies and the methods employed (which probably also requires new approaches and a lot more computing power than is currently available), both the first and fourth limitations relate to the deeper philosophical issues that have occupied researchers for a long time and should not be expected to vanish any time soon.

8.2 Summary and Outlook

Now, what have we achieved over the course of the last seven chapters? The book has addressed the overall research question: *(How) can memetics be fruitfully utilized for evolutionary economics?* This question has been divided into the following four sub-questions:

- *Q1: (How) can a memetic approach to economic evolution help to reveal links and build bridges between different but complementary concepts and approaches in evolutionary economics and related disciplines?*
- *Q2: (How) can the diversity and interdependence of characteristic elements of an organizational culture be captured such that these elements can be considered as representations of the underlying organizational memes?*
- *Q3: How can we model the diffusion of knowledge in innovation networks while taking a memetic representation of knowledge seriously?*
- $Q4_a$: *Which endogenous (meme-centered) elements of the Ice Bucket Challenge meme contributed to its diffusion pattern?* and $Q4_b$: *Can we identify structural factors (i.e., network properties) that influenced the diffusion of the Ice Bucket Challenge meme on social networks?*

The key findings of this book can be summarized as follows. Regarding *Q1*, I have suggested in Chap. 3 that not all definitions of memes are equally suitable but that an "informationalist" approach can link to various concepts in evolutionary economics, especially complex population systems, imitation heuristics, habits, and the rule-based approach. While I caution against exaggerated expectations about the explanatory potential of memetics *alone* with regard to economic evolution, I have also argued that the meme's eye view is a valuable perspective that should not be dismissed prematurely by evolutionary economists.

From Chap. 4, where *Q2* is addressed, we have learned that the complexity of organizational culture—in line with the notion of a complex population system of memes—can be captured by means of a memetic approach that draws on network science and a meme mapping technique known from marketing literature. Moreover, in Chap. 4, the first state-of-the-art review of the organizational memetics literature has been presented, thereby revealing much potential for further work in this area— especially with regard to applying the meme mapping approach and network analysis to the study of organizational culture. More precisely, we have seen how we can measure the differences of organizational memes in terms of interconnectedness and relevance. While the study presented in Chap. 4 should only be seen as a small step in the direction of a network science approach to organizational memetics, the results already indicate that both researchers and managers should pay more attention to the complexity of organizational memes.

With regard to *Q3*, Chap. 5 has taken the implications from Chap. 3 about the informationalist perspective on memes seriously by presenting a novel agent-based simulating model of compatibility-based knowledge diffusion and assimilation in innovation networks. The model developed in Chap. 5 takes several insights from memetics and other related disciplines into account. More precisely, the model can show how knowledge diffusion changes once we shift from the previous (e.g., vector-based) representations of knowledge to a network representation that also accounts for limited attention (and thus selection) as well as compatibility-based diffusion (and thus biased transmission and knowledge recombination). The implications of this new way of modeling knowledge diffusion and assimilation are manifold: For example, the results of the current model already contribute to contemporary debates about optimal network structures, which is an important issue in innovation policy. More specifically, the distribution of agent or node characteristics on an innovation network may often be more important for social learning processes than the particular structure or topology of the network itself. In this regard, future research could take this finding as a starting point because the distribution of characteristics may indeed self-organize within dynamic networks, whereas this self-organization can be assumed to strongly depend on variation, selection, and retention processes *within* the individual agents. This has, so far, remained under-researched by evolutionary economists. Here, the model can also serve as a basis for future studies that incorporate novelty creation by means of variation and recombination of existing knowledge units (or remix). Additionally, we have contributed to the literature on exploitation and exploration strategies in organizational learning by illustrating the performance differences of the two strategies depending on various levels of knowledge diversity between and within the actors in the network.

Finally, both parts of *Q4* have been tackled in Chap. 6 by using a two-pronged approach consisting of a descriptive and a computational part. First, the suitability of a memetic perspective for analyzing the diffusion pattern of the Ice Bucket Challenge has been demonstrated by means of an analysis that takes up insights about the actual diffusion pattern especially from Twitter and Google as well as theoretical propositions from the contemporary literature on criteria for successful replication. Second, the importance of structural properties of the network and prestigious individuals in the network (celebrities) have been demonstrated by means of a simulation model. Taken as a whole, Chap. 6 has contributed to a better understanding of "viral" challenges that can also be taken up by researchers, managers, and policy-makers that are interested in ways to influence the spread of such viral phenomena.

I would now like to propose a summarizing statement that may help to remember the core messages of my approach to econometrics. So, here is what I would call my *five i*: **Imitation** plays an important role for cultural and economic evolution, but memes are not just units of imitation but also units of **information** that contain rule-based elements of **instruction**, which—through processes of variation, selection, and retention—can lead to **innovation**, which does not occur in isolation but emerges from **interconnection**, that is, in networks and complex systems.

While all four studies presented in Chaps. 3–6 already contain specific suggestions for future research that will not be repeated here, the discussions in the previous Chap. 7 and the book as a whole have revealed several general directions that could be pursued in future research: First, the philosophical and also sociological implications of econememetics should be pursued further with regard to agency, creativity, normativity, and philosophy of mind. Second, the discussions have revealed that there appear to be various overlaps between memetics, *framing* (e.g., Lakoff 2014; Spitzberg 2014; Waddock 2019), and the study of signs and sign processes, that is, *semiotics* (see also Fomin, 2019; Kilpinen 2008; Botz-Bornstein, 2008; Juming 2009; Nye 2011; Lissack 2003), which can be systematically explored in future research endeavors. A third promising line of research could delve deeper into the significance and different forms of imitation, for example, by taking up insights from organizational institutionalism (especially the concept of *mimetic isomorphism*; e.g., DiMaggio and Powell 1983) or the literature on mimesis and culture (e.g., Garrels 2011; Girard et al., 2007; Palaver 2013). A fourth set of connections to other relevant (and "compatible") economic disciplines and approaches that could be pursued further includes *complexity economics* (e.g., Arthur 2015; Wilson and Kirman 2016), the *economics of attention* (e.g., Bernardy 2014; Davenport and Beck 2001; Falkinger 2007, Falkinger 2008; Franck 1998, 1999, 2005; Nolte 2005), and *bioeconomics* as well as *ecological economics* (e.g., Corning 2005; Costanza et al., 2015; Georgescu-Roegen 1971; van den Bergh 2018; Waring 2010; Witt 1999). A fifth avenue for future research on econememetics should aim at improving our understanding of *information*. Especially since information is a central concept in the literature on cultural evolution and also in evolutionary economics (e.g., in the generalized Darwinian approach; Aldrich et al. 2008; Hodgson and Knudsen 2010), the notion of information itself calls for more research, particularly in connection with information transfer by imitation.

As we can see, while this book may not have "solved" many problems, it has built many bridges to related and relevant disciplines that can serve as a sound basis for solutions to complex problems connected to cultural and economic evolution in the future. It can even be argued that econememetics may serve as a so-called *interfield theory* (Darden and Maull, 1977) for evolutionary economics.[1] Hence, to directly address the overarching research question: There are various promising areas where memetics can be fruitfully utilized but there is still much more work to be done.

Finally, to conclude this book, the importance of network science for future studies on econememetics cannot be stressed enough. If the core message of this book had to be reduced to one single phrase, I would probably go with "no meme is an island" (Dennett 1995, p. 144).[2]

References

Abatecola, G., Breslin, D., & Kask, J. (2020). Do organizations really co-evolve? Problematizing co-evolutionary change in management and organization studies. *Technological Forecasting and Social Change, 155*, 119964. https://doi.org/10.1016/j.techfore.2020.119964.

---

[1] See also the discussions by Kronfeldner (2009, 2011) in this regard.
[2] Alternatively, we could borrow from the title of my colleague Kristina Bogner's (2019) dissertation: "United we stand, divided we fall."

Aldrich, H. E., Hodgson, G. M., Hull, D. L., Knudsen, T., Mokyr, J., & Vanberg, V. J. (2008). In defence of generalized Darwinism. *Journal of Evolutionary Economics, 18*, 577–596. https://doi.org/10.1007/s00191-008-0110-z.

Almudi, I., & Fatas-Villafranca, F. (2018). Promotion and coevolutionary dynamics in contemporary capitalism. *Journal of Economic Issues, 52*(1), 80–102. https://doi.org/10.1080/00213624.2018.1430943.

Arthur, W. B. (2015). *Complexity and the economy*. Oxford: Oxford University Press.

Baldwin, J. R., Faulkner, S. L., Hecht, M. L., & Lindsley, S. L. (Eds.). (2006). *Redefining culture: Perspectives across the disciplines*. Mahwah: Lawrence Erlbaum Associates.

Bernardy, J. (2014). *Aufmerksamkeit als Kapital: Formen des mentalen Kapitalismus*. Marburg: Tectum.

Bogner, K. (2019). *United we stand, divided we fall: Essays on knowledge and its diffusion in innovation networks*. Doctoral dissertation, University of Hohenheim, Stuttgart. Retrieved from http://nbn-resolving.de/urn:nbn:de:bsz:100-opus-16151.

Botz-Bornstein, T. (2008). Can memes play games? Memetics and the problem of space. In T. Botz-Bornstein (Ed.), *Culture, nature, memes* (pp. 142–157). Newcastle upon Tyne: Cambridge Scholars Publishing.

Breslin, D. (2016). What evolves in organizational co-evolution? *Journal of Management & Governance, 20*(1), 45–67. https://doi.org/10.1007/s10997-014-9302-0.

Corning, P. A. (2005). *Holistic Darwinism: Synergy, cybernetics, and the bioeconomics of evolution*. Chicago: University of Chicago Press.

Costanza, R., Cumberland, J. H., Daly, H., Goodland, R., Norgaard, R. B., Kubiszewski, I., et al. (2015). *An introduction to ecological economics* (2nd ed.). Hoboken: CRC Press.

Darden, L., & Maull, N. (1977). Interfield theories. *Philosophy of Science, 44*(1), 43–64. https://doi.org/10.1086/288723.

Davenport, T. H., & Beck, J. C. (2001). *The attention economy: Understanding the new currency of business*. Boston: Harvard Business School Press.

Dennett, D. C. (1995). *Darwin's dangerous idea: Evolution and the meanings of life*. New York: Simon & Schuster.

DiMaggio, P. J., & Powell, W. W. (1983). The iron cage revisited: Institutional isomorphism and collective rationality in organizational fields. *American Sociological Review, 48*(2), 147. https://doi.org/10.2307/2095101.

Durham, W. H. (1991). *Coevolution: Genes, culture, and human diversity*. Stanford: Stanford University Press.

Falkinger, J. (2007). Attention economies. *Journal of Economic Theory, 133*, 266–294. https://doi.org/10.1016/j.jet.2005.12.001.

Falkinger, J. (2008). Limited attention as a scarce resource in information-rich economies. *The Economic Journal, 118*(532), 1596–1620. https://doi.org/10.1111/j.1468-0297.2008.02182.x.

Fomin, I. (2019). Memes, genes, and signs: Semiotics in the conceptual interface of evolutionary biology and memetics. *Semiotica, 230*. https://doi.org/10.1515/sem-2018-0016.

Franck, G. (1998). *Ökonomie der Aufmerksamkeit*. Munich: Hanser.

Franck, G. (1999). Jenseits von Geld und Information: Zur Ökonomie der Aufmerksamkeit. *Kunstforum International, 148*(Dec 1999–Jan 2000), 84–94.

Franck, G. (2005). *Mentaler Kapitalismus: Eine politische Ökonomie des Geistes*. Munich: Hanser.

Gallie, W. B. (1956). Essentially contested concepts. *Proceedings of the Aristotelian Society, 56*, 167–198. https://doi.org/10.1093/aristotelian/56.1.167.

Garrels, S. R. (Ed.). (2011). *Mimesis and science: Empirical research on imitation and the mimetic theory of culture and religion*. East Lansing: Michigan State University Press.

Georgescu-Roegen, N. (1971). *The entropy law and the economic process*. Cambridge: Harvard University Press.

Girard, R., Antonello, P., Rocha, J. C. d. C. (2007). *Evolution and conversion: Dialogues on the origins of culture*. London: Continuum.

Hodgson, G. M., & Knudsen, T. (2010). *Darwin's conjecture: The search for general principles of social and economic evolution*. Chicago: University of Chicago Press.

Illies, C. (2010). Biologie statt Philosophie? Evolutionäre Kulturerklärungen und ihre Grenzen. In V. Gerhardt & J. Nida-Rümelin (Eds.), *Evolution in Natur und Kultur* (pp. 15-38). Berlin: de Gruyter.

Juming, S. (2009). The magic of meme—On memetics and its development in China. *Chinese Semiotic Studies, 2*(1). https://doi.org/10.1515/css-2009-0110

Kilpinen, E. (2008). Memes versus signs: On the use of meaning concepts about nature and culture. *Semiotica, 2008*(171), 305. https://doi.org/10.1515/SEMI.2008.075.

Kroeber, A. L., & Kluckhohn, C. (1952). *Culture: A critical review of concepts and definitions*. Cambridge: Peabody Museum of American Archaeology and Ethnology, Harvard University.

Kronfeldner, M. (2011). *Darwinian creativity and memetics*. Durham: Acumen.

Kronfeldner, M. E. (2009). Meme, Meme, Meme: Darwins Erben und die Kultur. *Philosophia Naturalis, 46*(1), 36–60.

Lakoff, G. (2014). *The all new don't think of an elephant! Know your values and frame the debate*. White River Junction: Chelsea Green Publishing.

Lissack, M. R. (2003). The redefinition of memes: Ascribing meaning to an empty cliché. *Emergence, 5*(3), 48–56. https://doi.org/10.1207/s15327000em0503_6.

Nolte, K. (2005). *Der Kampf um Aufmerksamkeit: Wie Medien, Wirtschaft und Politik um eine knappe Ressource ringen*. Frankfurt a. M.: Campus.

Nye, B. D. (2011). Modeling memes: A memetic view of affordance learning. *Publicly accessible Penn Dissertations. 336*. Retrieved from http://repository.upenn.edu/edissertations/336/

Palaver, W. (2013). *René Girard's mimetic theory*. East Lansing: Michigan State University Press.

Rosenberg, A. (2017). Why social science is biological science. *Journal for General Philosophy of Science, 48*(3), 341–369. https://doi.org/10.1007/s10838-017-9365-0.

Spitzberg, B. H. (2014). Toward a model of meme diffusion (M^3D). *Communication Theory, 24*(3), 311–339. https://doi.org/10.1111/comt.12042.

van den Bergh, J. C. J. M. (2018). *Human evolution beyond biology and culture: Evolutionary social, environmental and policy sciences*. Cambridge: Cambridge University Press.

Waddock, S. (2019). Shaping the shift: Shamanic leadership, memes, and transformation. *Journal of Business Ethics, 155*(4), 931–939. https://doi.org/10.1007/s10551-018-3900-8.

Waring, T. M. (2010). New evolutionary foundations: Theoretical requirements for a science of sustainability. *Ecological Economics, 69*(4), 718–730. https://doi.org/10.1016/j.ecolecon.2008.10.017.

Wilson, D. S., & Kirman, A. P. (Eds.). (2016). *Complexity and evolution: Toward a new synthesis for economics*. Cambridge: The MIT Press.

Witt, U. (1999). Bioeconomics as economics from a Darwinian perspective. *Journal of Bioeconomics, 1*(1), 19–34. https://doi.org/10.1023/A:1010054006102.

Correction to: "Meme Wars": A Brief Overview of Memetics and Some Essential Context

Michael P. Schlaile

Correction to:
Chapter 2 in: M. P. Schlaile (ed.),
Memetics and Evolutionary Economics, Economic Complexity and Evolution,
https://doi.org/10.1007/978-3-030-59955-3_2

In the original version of the book, In Chapter 2, the word "muchlonger" in page 15 is now corrected as "much longer" and the word "several doctoraldissertations" in page 17 is now corrected as "several doctoral dissertations". The chapter and book have been updated with the change.

The updated version of this chapter can be found at
https://doi.org/10.1007/978-3-030-59955-3_2

© The Author(s), under exclusive license to Springer Nature Switzerland AG 2021
M. P. Schlaile (ed.), *Memetics and Evolutionary Economics*, Economic Complexity and Evolution, https://doi.org/10.1007/978-3-030-59955-3_9

Manufactured by Amazon.ca
Bolton, ON